QUANTUM RELIGION

THE GOOD NEWS OF RISING CONSCIOUSNESS

BOOK TWO

OF THE "SECOND ENLIGHTENMENT" TRILOGY

By

Sylvester L. Steffen

ISBN: 1-4107-3975-9 (e-book)
ISBN: 1-4107-3977-5 (Paperback)
ISBN: 1-4107-3976-7 (Dust Jacket)

Library of Congress Control Number: 2003092482

This book is printed on acid free paper.

Printed in the United States of America
Bloomington, IN

The cover graphic of the Sun, Moon and stars was computer-drawn by Monica R. Steffen.

1stBooks – rev. 10/20/04

The Second Enlightenment

The Thirty Years Wars of Religions (1618-1648) left Europe wasted, religiously, politically, and economically. In the centuries preceding the wars, multivalent civil and religious tensions quaked across Europe and gradually gathered an epochal storm with many tornadic centers. By 1648 the energy of the storm was spent, and the sun re-appeared over a Europe in shambles. In the Peace of Westphalia the warring parties acknowledged that Europe would not be a united Holy Roman Empire under a joined civil and religious authority as it had been. Reluctantly but necessarily, the parties signed on to the peace, even though they held on to their animosities and professed absolutisms.

The beginning of the "first enlightenment" is the beginning of "modernity" which deconstructed the mythologies that energized the warring absolutisms. The *second enlightenment* now emerging is an insight recognizing value in the naturally evolved diversifications of religious, political, and cultural systems, and, in the embrace of differences, not the mere grudging tolerance of them. The "second enlightenment" is a gathering postmodern vision of better ways for social reconstruction, for example, of the recognition that there is greater benefit in the symbiotic potential of religious, political, and cultural collaboration than in their unaccommodating opposition.

Out of historical continuity sense can be made of *the signs of the times*. From the seventeenth century's Wars of Religions came painful apocalypse and enlightenment. From the World Wars of the twentieth century also come "un-coverings" (apocalypse) and enlightenment. New Prophets, the Statesman Mohandas K. Gandhi and Churchman Martin Luther King, witness against war's futility and for the worthy purposes of non-violent activism. The lethality of modern war weapons threatens ultimate waste.

The Second Vatican Council is a compelling spiritual event, whose transparency stands in stark contrast to the institutional-serving cover-ups enabled by the Councils of Trent and the First Vatican.

Through the overcast of human obduracy shines hope for a *second enlightenment* of evolved rationality—the wisdom continuity of cosmic evolution and the pulsating soul of the expanding universe. Cosmic rationality is the patiently rising insight of global consciousness; rationality is the universal well of consciousness divined in the evolving in-tension and intention of quantum-electric spirituality and materiality. The light from within, the cosmic light of common rationality may yet illumine for humans their natural place of divine grace.

Remembering

Monica Ruth Steffen
April 5, 1960—June 9, 2001

SELF PORTRAIT, circa April 2000

Softly, as the morning dew she came, and went.
She took upon herself personally Earth's causes
And on Earth's behalf she suffered willingly.
Hers was the pathos of divinity, the smile of an angel,
The solicitude of a mother and the soft touch of a grandmother.
She had the whit of wisdom and
The flinty conscience of a prophet.
She was honest beyond pretense.

Like the scent of incense her blessings linger.
Her candle illumines ever. Her flame dims never.
Her company is with the saints.
Her deepest solicitude she reserved for the children.

In the face of death she knew how to live.
She etched with faltering hand her final Scripture:
"It's a fearful thing to love what death can touch".

Monica Ruth Steffen

Perhaps the most striking and memorable quality Monica impressed upon others during her life was her artistic talent. She had the unique ability to recognize the true beauty in anything around her—whether a tree, cat, plant, painting, flower, rock, building or person. Monica had the gift of seeing everything in an aesthetic way—but her recognition of the beauty in life went far beyond traditional concepts of what is beautiful. She could see beyond the physical qualities of the things around her and tap into the true inner beauty that underlies everything.

Her artistic perceptions translated into the way she lived her life, expressed her self and developed her own unique style of art. Her watercolors, sculptures, sketches, carvings, music, singing, etc., all reflected a beauty that so often transcended the ordinary. In addition, Monica's willingness to help others—whether they needed assistance with an art project, advice on a landscape design, support in dealing with a personal problem or encouragement to develop their individual skills—revealed just how important sharing her unique perspective on life was to her.

Monica's life was always about giving and sharing. She gave of herself—be it her talents, her time, her money or her advice—freely and happily. People who received her gifts recognized immediately that Monica possessed a very unique gift; each of these people benefited immensely from Monica's sharing.

While all who loved her in life will forever miss her presence and her talents, we know that the gifts she left have made her immortal. In donating her eyes, Monica's unique way of seeing the beauty of life lives on.

Leticia L. Steffen

Author's Foreword
Religion, or Theology?

WHY CALL THIS BOOK Quantum *Religion* and not Quantum *Theology*? For the simple reason that religion has everything to do with relationships, with action, with work. Theology is really a *philosophy of religion*, the *God word* behind religion. Sometimes it seems there are more words than action, like—less than an ounce of action (work) for every pound of words. We live religion and talk theology. Work and word are complementary and necessary; they are together the experience of Sacrament.

RELIGION PERTAINS TO RELATIONSHIP, to the common consciousness of essential interdependency that determines for our lifetimes our spiritual and material well-being. Religion/relationship is a consciousness that bears directly on daily living from birth to death. Like consciousness, religion/relationship grows as we age in experience, wisdom and grace. Consciousness is an ever-rising awareness of self that comes to us with communication in all its spiritual and material dimensions. The faculty of consciousness reciprocates in kind according to childhood experiences. Everybody communicates, and the development of communication skill is fascinating to experience and to observe. Communication is a process of direct "information" exchange, which, as such, goes before speculation and presumption. The word "religion" is mostly understood as an instituted body of people who corporately hold to particular beliefs and practices developed from a particular strain of theology. Instituted religion is called "Church". *Religion* pertains to ways and means of people in their daily, common efforts to understand and live in right relationships and for common well-being. In a personal manner, each is called to "do" religion and to "do" theology. *Doing religion* is in essence a matter of doing life by informed conscience, which, it should be noted, obliges also those who choose to do theology professionally for and in the church institution. Simply stated, true religion includes, as a matter of elemental consideration, love and respect (hands-on care) for nature—

parent to all of us—even as we love and respect our parents and one another. This root consciousness authentically inspires religion and theology.

THEOLOGY IS THE ART OF SPECULATIVE REASON, a derivative process based on assumptions and speculation; these concern matters of creation, personal existence, personal and communal relationships, the nature of life, suffering, death, etc. Theology is an evolved body of information, typically developed by *dialectics*, the exercise of logic, rational argumentation, information-based assumptions and speculation. Thus, theology is more a product of ongoing mental gymnastics and construction than is religion (relationship). Not everyone is equally drawn to or compelled by high theological speculation for it is a professionally disciplined "art-form" that is more or less credible according to the credibility of its creeds and doctrines, its casuistry, and the assumptions underlying them. Many consider theology and its disciplined process to be esoteric, and not always credible because old assumptions, presumptions and speculations do not always keep pace with changing knowledge. Loss of theological credibility bears directly on the institutional churches of today.

THE FAITHFUL RIGHTLY HOLDS CHURCH to be rightly informed. When Church membership accepts new fact-knowledge that requires the updating of theological thinking, the membership expects its theologians to correct out-dated theology premised in misinformed "facts". It seems obvious that if theology has its premises wrong, then, conclusions based on them will go off in misinformed directions and will misdirect. And, so it has happened and continues to happen. Fixations in fact errors become pretexts of religious contention, wars, and internecine violence.

WHILE THIS BOOK DEALS WITH FIXATIONS of traditional theologies in fact errors, it deals more importantly with religion, with relationships and contemporary understandings of life's essential continuities and "providential" dependencies that uplift consciousness and wholesome living. Thus, while this work seeks to inform in regards to updated fact-knowledge, its real intention is to motivate public wholesomeness and right-mindedness in daily, lived relationships, and to contribute to a public consciousness of essential grounding in "universal" relationships that facilitate the uplift of consciousness.

WHEN THE PRESENCE OF ONE SPECIES becomes so prolific that it dominates all other life forms, it causes nature to lose its capacity to maintain a sustainable balance for all species. Obviously, the excessively dominant species must eventually suffer reversals when the fabric of life in which it thrives is catastrophically shredded. Professionals and non-professionals now believe that a critical state of resource consumption and ecological degradation has been reached and is tipping the scales of balance toward out-of-control network collapses. It is now scientifically established that human exploitation of reef habitats and trees, for example, is impacting climate changes and water distribution with the long term and worsening degradation of land-and-sea dependent ecologies. *Doing theology* as usual seems, under the circumstances, to be irrelevant so long as theology continues to operate conscionably disconnected from the authentication of natural necessity. The collective destruction of network life is real abortion that flies in the face of religion and theology alike; activism against individual abortion is no saving salve for the abortions to which culture is blind.

RISING CONSCIOUSNESS IS GOOD NEWS because it introduces us ever more profoundly into the unfolding mysteries of ongoing creation, of the self, of God and new conscience. If the "*salvation* (the fulfillment of personal/social destinies) *message*" of Jesus is true, and the good sense of faith tells us that it is, we will discover, in delving into cosmic rationality, the spiritual/material certitude of our faith and the grounding of the communal persona in Creation Theology—the ongoing revelation of ascendant rationality/purpose.

TRANSPARENCY OF INTENTION SHOULD SHINE through all words, whether the words operate on the materially obvious plane or at below-the-surface levels of deep consciousness. The reader will discover that this writer struggles through the circuitous conventions of theological niceties, but also, mercifully, he at times manages his way through things less obscure; in either case, it is his prayer that the truths of common sense surface over all, for if this isn't accomplished—the test of intellectual honesty—then he also prostitutes words.

It should be realized that the traditional consciousness of institutional religion arises from the common well of consciousness—*cosmic rationality*. If this is true, it implies then that to the extent cultural ideologies are faithful to universally authenticated rationality, they will, in their outcomes, prove to be valid; on the other hand, to the extent that cultures diverge from universally validated rationality, they will be exposed for their invalidity. Invalidity registers in corrupting outcomes, such as violence and ecological desecration. Societal fixations in staticism, centrism and absolutism, whether they occur in Judaic, Christian, Islamic or other religions/cultures, violate cosmic rationality and fail the historical test of validity, credibility and humanity. Reason, conscience, requires that invalidity in relationships be contested and eliminated.

Quantum religion **is a new consciousness** unfolding a bit at a time in the historical context of evolving cultures. Instead of fixating in a suppressive and overwrought deity-consciousness, as has been customary in institutional religions, *quantum religion* accepts that divinity self-reveals constantly, personally, communally, and instant-by-instant, in a manner sufficient to secure personal and societal welfare. *Quantum Religion* presupposes that consciousness will remain conscionably engaged in communication.

A challenge to individual person and to civilization is the recognition of the *quantum instance of divinity*. This awareness is the *hypostasis* of all "understanding" of "natural connection" in the human and the divine. The recognition and acceptance of divine presence *in the least quantum* is authentic to universal religion for it is a consciousness that is faithful to cosmic continuity and rationality. Religion begins *intuitively*, as the wisdom/conviction of understanding/faith, and matures personally to *reflectivity*, to the individual awareness and acceptance of the communal, conscionable duty of providing for common sustainability and welfare.

Quantum religion is the conscionable continuity of evolving rationality, which advances on the three-stage processes of communication, consciousness, and conscience. In its conscionable ascendancy, *quantum religion* requires all and each to do as Jesus did—intentionally conform personal conduct to communal necessity and to sustainable well-being.

PERHAPS FOR THE FIRST TIME IN HUMAN HISTORY

GLOBAL CONSCIOUSNESS IS OPEN TO

QUANTUM RELIGION

Table of Contents

SECTION II

SECTION III

SECTION I

ON REASON

REASON is used mostly in two senses, as *agency* (faculty) and as *process*. As agency, "reason" causes things to happen; it is a cause-and-effect *agent* (doer). For example, some would say that *laziness is the cause of lethargy*.

And, "reason" is used to mean "the faculty of intelligence"; for example, cumulative knowledge and experience increase reason's capacity for understanding. Cause-and-effect logic tells us that nothing happens without a cause; and because our self-aware consciousness is *intuitively reasonable* it wants to know causes. This faculty of consciousness, wanting to know causes, is "rationality", the intelligence of reason.

Based on these understandings of reason, two definitions can be advanced: 1.) reason is the active **faculty** of conscious intelligence that fulfills self by engaging the *process of reason*, namely, the spiritual (energetic) processes of communication, consciousness and conscience; and, 2.) "reason" is a social **process** of intelligence that enables intentional judgment, individually and collectively, to make choices that benefit personal and social well-being.

For many reasons, it is clear that *as agent trying to look into its own self's agency* we cannot know the physical/psychical train of causal relationships that makes reasoning work in our faculty of intelligence, nevertheless, we do have, from cause-and-effect experience, intuitional certitude, by which we "know" and can anticipate as to likely cause(s) and specific effects.

Cumulative "science", the acquired knowledge of the evolutionary make-up of complex nature (energy/matter), gives humankind a common sense of the "quantum-relative", namely, that every least thing is *related* to every other thing, and of the cause-and-effect train of events and relationships—the evolution of energy/matter. Thus, the exercise of the faculty of reason by the engagement of its processes, (communication, consciousness and conscience), is with personal and social consequence, including the enlargement of the collective rational faculty in its experience of and capacity for faith, hope and love.

It is the "Grail Quest" of reason that drives consciousness to penetrate ever more deeply into the valuation of everything—including our selves—in *quantum relative* and *evolutionary* terms, in order to gain self-knowledge and discover "ultimate" causes (origins) and effects (destinies).

As *faculty* (structure, matter) and *process* (energy, spirit), reason is essentially "natural sacrament". The pattern is Sacrament—the nature/nurture of sign/grace. Reason is nature and nurture. Reason is sign and grace. Reason is embodied consciousness. Self-conscious rationality is a high subtlety of cosmic transformation. Reason is structure of energetic logic. Authentic reason transforms authentically; the converse is true also.

Not only is there rational causality in nature, nature is rationality. Ignorance and arrogance compromise authenticity, rationality. Reason is purpose; purposelessness is irrationality. Only humans can choose to be irrational. Sustainable transformation is essentially rational, essentially purposeful. Irrationality scrambles cosmic consciousness. Children, generation after generation, are victimized by cultured irrationality and are prematurely and lastingly frustrated in rational potential. This has to change, and it can.

Cosmic Rationality

Traditional Western religious culture has engrained in us an understanding of conscious soul as life's agency, separable and distinct from the body. In the personally lived experience of the individual the evidence is of soul/body unity. Even more, we are cultured in the perception that our material bodies, their appetites, indeed, materiality itself, are enemies to spirituality, to *salvation*. Salvation (becoming whole) is spiritual and material healthiness (*sanitas*—sanity). We have yet to awaken to the reality that spirituality and materiality are coincident, codependent, and unified in self-identity. We have yet to wake-up to the spiritual nature of physical reality and recognize that human self-awareness is an extension and refinement of cosmic consciousness, embodied consciousness, the necessary, communicational agency of all continuously embodied spirituality.

One might ask, "What are the cosmic forces of transformation and evolution?" They are the evolved continuity-energies of the Big Bang, *inherency* and *coherency*. Inherency may be understood as the unique internal energy of discrete cosmic particles—the wave/particle energy of quantum-electric potential. Coherency may be understood as the continuity bonding of complexities derived from particulate inherency—Einstein's C^2, the speed-of-light constant of his Special Relativity equation. Coherence accounts for the continuity of complexity, which enfolds the conscious potentials of energetic openness. Spirituality is the qualification of coherency, as is the rational resonance of communication, consciousness and conscience. The cohesive intension of atoms and molecules is the continuity force of transformation and evolution.

Atomic/molecular *intension* has evolved into the self-aware subtlety of *intention*. While intention is an evolved complexity of intension, it differs from intension in that, unlike intension, which is *reflexive (ex opere operato)*, driven by natural laws of cosmic energy/matter, intention is *reflective (ex opere operantis)*, a self-awareness that can choose to act or not to act depending upon

4

anticipated outcomes of action taken or not taken. The intension in atomic/molecular action/reaction is physically determined by properties of the atomic/molecular constituents. Intension and intention are kinds of consciousness belonging as essential properties (spirituality) to embodied cosmic rationality.

Cosmic rationality is the whole of universal (universe) intelligence, the energetic, conscious spirituality of all cosmic mass (MC^2). It is located in the diverse materiality of all cosmic substance, distributed in and throughout the universe. It is open potential in the universal ocean of cosmic plasma. The "unity of intelligence" is premised on and rooted in the energy/matter continuity, produced, and still being produced, by Big-Bang expansion. Also, qualified energy, consciousness, is included in Einstein's "E".

At some time "early" after first Big Bang expansion began, energy waves came to be tensioned in particulate (string) dimension. Dimension-conformed energy diversifies into polar potentialities based on positive/negative electrical charges. On the bases of energetic compatibility, particulate differentiation included the subtle complexes of joined particles, which are actively engaged according to positive/negative attractions, aspects of Einstein's Theory of General Relativity. Electromagnetism, differentiated quantum-electrically and transmitted in wave pulses, is an elemental aspect of essential gravitation, whose amplified energy is molecularly accessed by attenuation.

Complex particulate differentiation weaves the subsidiary bases for involving energy modifications in subtle and controlled interactions.

The popular Big Bang Theory holds that the cosmos began in a violent outburst of pent-up power, which sources the elemental potentials for violence and for tensioned order—syntropy. Subtle complexities represent the containing of violence and the introduction of sustainability as an accommodation of tensioned energy, conformed and diversified. Conflicts ferment at the poles of cumulative electrical potential, where the pooled charges foment toward liaisons with opposite charges pooled at some other pole. Great energy exchange occurs when electrical charge bridges across the polar divide and connects with pooled, opposite charges of another pole. Creative potential is discovered and expressed in the connection of unlike potentials accumulated in particulate links. On the subtleties of particulate expression creation unfolds, and infinite varieties of the energetic/substantive appear. New diversifications create other new potentials for further creative communications. Communication is the common mechanism giving definition to context, content and connection for the simplest and the most complex cosmic outcomes. All outcomes, all diversification, all "rationality", begin with communication, whose outcomes include attraction, repulsion, mixing, disassociation, association, convergence, divergence, expression, accommodation, connection, transmission, transformation and amplification.

The outcomes of communication are understood to be "conscious", the expressed products of amassed, particulate energy. Particulate forms have manifestations, which qualify "matter", while the potentials for transformation qualify the expressions of energy (wave transmissions). All expressions of energy/matter are *noumenal* (spiritual) and *phenomenal* (material), including the self-aware human person. It is axiomatic, that is, *self-evident*, that the continuity of cosmic expressions evolved from/in Big Bang expansion should manifest consistency and certain conformity (order); cosmic energy (*virtue*) and its embodiment, structure, and form (*sign)* are universal sacrament. Big Bang potential corresponds noumenally and phenomenally to evolutionary outcomes. God's voice is Nature's. Creator God is subtly "ineffable" in evolutionary creation.

An aside: recognition of this mystery may facilitate a resolution to the dogged debate between creationism and evolution. Perhaps the discussion should be framed in a way that does not require an "either/or" solution, but should be open to a "both/and" solution. The both/and resolution allows for codependent continuity between evolution and intelligent-plan in creation. Evolution is the sign/grace, the nature/nurture of God Present. Transformation (creative outcome) contains the grace represented in the sign. The "both/and" principle, applied to creation and evolution, is applicable also to "determinism" and "free will". These terms are not mutually exclusive polarities; like nature/nurture, they together drive the interactive forces of creation.

6

"Rationality" pertains to the conscionable awareness aspect of subtly *evolved energy*, the noumenal, spiritual aspect of substantive (phenomenal) characterization in all its material variations. The continuity of evolved substantiation involves a rationality of purposeful outcomes, which is to say, it involves the *sustainability of liaisons* that are successfully, successively, and repetitively self-reproduced. From the cosmic perspective, that is, from the self-aware introspection of embodied spirituality, purpose is experienced *a priori* and *a posteriori* as *cause* of and *outcome* of evolutionary success. Rationality is the evolved body of subtle consciousness, and is the process (dynamic logic) by which new consciousness arises. Thus, it is accurate to say that rationality, as outcome and process, is a cosmic necessity in which humankind, body and soul, is essentially implicated. The implicate expectations of transformation derive from cosmic intension and intention.

Religion (theology) that fails to take into consideration the necessities of cosmic rationality fails in its essential task. The *intuitive* and the *reflective* aspects of rationality are both weighed in St. Anselm's weighty phrase "fides quaerens intellectum". The prior (all things unknown) consciousness, reason, intelligence, is deep intuition, which gives to faith it's grounding and credibility. Faith grounds in the sustainable experience of prior rationality. Memory holds sacred and reiterates experiences of sustainable rationality in correlated and coevolved qualifications. Insights into memory validate faith.

The rationality of experience is intuitionally bonded in embodied consciousness. Rationality and faith are processed in the same neural complexes. Faith's grounding is enmeshed in the rationality of experienced intelligence. Thus, faith is a product and outcome of cosmic rationality, a priori and a fortiori. It is a misdirected and unfounded presumption to put faith in an independently prior relationship to intelligence, rationality. The perversion, which sets faith above and apart from rationality, is called "fideism", a hyper fixated tradition of belief in presumptions based on the mysterious unknown. The misdirection of handed-down religious orthodoxies (*foundationalism*, fundamentalism) consists in is: that the un-accommodated, literal belief in ancient scriptures objectifies, "statifies", and absolutizes the rationality of one time and remains closed to the ongoing revelations of reason that enable growth into new experiences of consciousness and understandings—the source of faith's continued revitalization. Fideistic objectification and fixation desecrate the *sacrament* power (*virtus*) of intelligence to uplift consciousness and bring rationality to new insights of conscience.

Belief in a Creator of the universe leads to belief that *in Creation the Creator is revealed.* Thus, the believer chooses to "read" the universe as the "primary revelation of the divine" (in Thomas Berry's words). In a revelatory and real way, the cosmos is *primary scripture,* the word/work of divine composition, and the continuity expression of the ineffable God. Human history is understood to reveal God-consciousness. Human personality and communality are understood as cosmic windows of consciousness into the Godhead personality.

If we live long enough we may find that *his*-tory/*her*-story is divisible into three parts (personalities), analogous to Trinity. In our early lives we experience "second person" relationships, the relationship between child and parents. In this relationship we experience a sharing of personality with our parents who have given to us our genetic make-up and who will invest a good portion of their lives in nurturing our body/soul by provisioning for us, by their example of living, and by the knowledge they impart to us. In image and likeness we carry for our lifetimes the family character given to us by our parents. In the family setting we learn the necessity of reciprocal relationships, that is, being not just the recipient of the beneficence of others, but being ourselves benefactors of love and service to others. In this period of life we grow into wisdom, age and grace.

Practical relationships in the *second person* experience prepare us for the matured, personal role of living, of ourselves being persons grown into the parenting role of the "first person". As male or female, the role of parenting has much in common, but also much that is unique in/to femaleness and maleness.

Scripture tells that humans are made to God's likeness, male and female. Sexual subtlety is the axiomatic expression of quantum-electric (*positive*, female, and *negative*, male) energy/matter. Femaleness, in symbol and fact, is characterized in quantum-electric grounding, while maleness is characterized in the negative energetic potential of free agent electrons swarming the nucleus. The high achievement of reflective consciousness in subtle sexuality represents a quantum leap in the advance of cosmic rationality. Based on evolutionary evidence, it seems right to credit sexuality with the leap in consciousness that we now experience in dialogic self-awareness. The left-right disposition of the body's limbs and organs testify to sexual polarity and mitotic cell structuring. Prior to the separation of female/male sexes, organisms reproduced by cell division, "asexually"; the function of reproduction by cell division is characteristically a *female* trademark. There are mechanisms in animal cells, including human, namely, the organelles, mitochondria and plastids, that have their own DNA, which is transferred cell-to-cell by cell division, and are carried in the female egg cytoplasm but not in the male sperm.

Mothers, by the endowments of *nature* and *nurture*, are the more stabilizing component of family for they are the persons in whom the nurturing of life begins and upon whom life most depends. Womanhood confers stability and grounding to family and society. Biologically, mothers invest personally the most in/to life. The ovum is the major determining matrix of individual life, male and female. In the cell matrix are all the raw materials of structural make-up needed in the growth of the *conceptus*. The ovum contributes three distinct strands of DNA; two pertain to essential cellular processes of energy maintenance and development (mitochondria and plastids), which are outside the ovum's nucleus. In the nucleus of the egg-cell, one-half the genetic chromosomes is provided, while the other half of the nuclear chromosomes is provided paternally, that is, by the male sperm. To have a graphic sense of the relative contributions of the ovum and the sperm, the sperm may be thought of as being the size of a tomato seed, and the tomato fruit as corresponding in size to the ovum; even more contrasting is the size of a rooster's sperm and a hen's egg.

The ovum is the ground state, the gravitational place of rest and purpose for the sperm, even as the mother is for her lifetime the place of refuge, solace and security for her conceptus, infant and child. For mother and child, the father is an enabling presence, who in word and work supports the mother's totally preoccupying task of nurture. In preserving the essential continuities of life, it is quite evident that the role of the woman is far more substantive (spiritually/materially) than is the role of the father. The mother's spirituality predominates naturally in conception and in the family setting, hence, it is right to analogize the role of her person in trinity as that of holy spirit. The mother contributes primarily, before birth and after, to the character of the spiritual dialogue occurring between parents, and between parents and children. Her spirit is essentially and necessarily "holy", if familial, communal harmony is to work. In the correspondence of purpose, father and children cooperate with the nurturing work of mother. It is in the communal relationships of mutual nurturing that the consciousness of Trinity is experienced and learned; this is true for family as it is for civil society. Anything that frustrates necessary family harmony registers also in the frustration of civil society. Familial trinity, dialog, is the heart and soul of religious, civil society.

The uniqueness and variability of female and male consciousness complement the diversification of conscious, human potential. There is no biological basis for the hyper arrogation of one sexual consciousness over the other, whether in familial or public authority. In the human species, the sexual division of femaleness and maleness enable rationality to put greater distance between mere molecular intension and the higher purpose of intention, the conscionable *hypostasis* (understanding) of willful harmony. The intentional motivation (hypostasis) of human relationships in the purposeful uplift of the *human/divine* person—word made flesh—is a quantum leap above the lesser consciousness of *stasis*.

Why bother about cosmic rationality? Because there is too much disconnection, too much social friction and violence prevailing in human relationships, aggravated by lack of informed insight. The improved understanding of cosmic/human rationality, our own, may facilitate generally better understandings and better relationships. All good sense, all common intelligence in the evolutionary continuum is included in "cosmic rationality". It is in the nature of intelligence to want to grow in good sense. We can never have too much good sense—but we all hurt from too little of it.

It is a deep-rooted hunger of intuition and conscious rationality to seek and to know the "Summum Bonum", the Ultimate Good in which all being originates and is sustained. Intuition provides the necessary appetite that hungers beyond the contrived divisions and ephemeral needs of changing truth. Intuition is bigger than the smallness of denominational, competitive religion. Intuition brooks no quarter for institutional divisiveness rather she seeks to bridge their divides by being inclusively open. She dares to spell her God with a small "g". Intuition is a consciousness questing beyond the confines of personal space and time; she attenuates the perennial energies of vital subtlety and amplifies her own for the edification of generations to come; she is the present and future ecstasy of resonance conformed to the harmonies of the spheres, small and large; she is the echo box of the violin that sings her full voice and brings each body cell to sound its soul's harmony; she is that meaning of heaven which conforms human consciousness to divine, and, which rises predictably like the morning Sun and Spring freshness; she is a spirituality that lives forever beyond the changing definitions of materiality; she is the wisdom of the ages; she is sensitively aware beyond hard instincts; she chooses not the harsh fixations that violate sensible refinements; she is the melding soul of love that purges from gold its dross; her zeal for truth is a double-edged sword whose acuity peals away the deceit of pretenders and lays bare truth's openness.

Intuition is a passion in each of us, which needs to be exercised. If she isn't exercised she is easily buried by preoccupation with other less worthy good. Our obsession with less permanent objects is less satisfying, for every object, like a crystal snowflake in the palm of our hand, disappears before our eyes as we cling to it. The glamour of things we obsess over loses its luster as they gather dust, and as they make demands on us greater than they deserve. What lasts are not things but relationships. The webs of mutual caring that we weave in a lifetime accumulate the greater value, in terms of the common good—more precious than the gold in Fort Knox. The work of weaving relationships may lack the glitter of gold, but the golden strings of light tethered in caring relationships are of infinitely greater value. In its use of metals in its vital weave, life perfects consciousness on ferrous foundations, which facilitate the cohesive agency of magnetic grounding. Blood hemoglobin and the molecular iron at its core contribute to the functional vitality of blood flowing in all animal veins to all parts of the body; they contribute *animus* to all animal communication, intension and intention.

The soul of a person compelled by love is that of a saint; her life is lived with the edged language of a biblical prophet who challenges greatness with smallness, blindness with vision; mindlessness with mindfulness; she indicts the sin of the mighty and extols the virtue of the small; with the sting of a bee she protects the heart of the rose and knows its sweet scent; and in the cause of justice she hesitates not to use her sting for life, even at personal risk.

In fairness to the reader I should acknowledge my limited competency to write on the topic of cosmic rationality. The only qualification I can offer to justify my daring is that I, like every other human being, personally embody cosmic rationality; we all belong to one continuum of original energy/matter derived from the common stuff of the Big Bang. In our own persons, we individually possess the intuitive awareness of all antiquity embedded in the neural mesh of our embodiment. Legitimate insights into common rationality are as multiple as the minds/bodies sharing common cosmic consciousness.

The word "cosmic" includes time/spatial complexities that are generally the specialty field of theoretical physicists. The word "rationality" pertains to primitive consciousness; in its basest and least state, it accounts for first communications in originally expanding energy/substance. More than primitive consciousness, rationality pertains to the continuity train of consciousness from the first instant of Big Bang expansion to the present time. The underlying assumption here is that cosmic substance, all life, all humankind, all consciousness is included in the one continuum that has evolved from first implications of energy/matter.

Rationality, in its capacity for self-conscious evaluation, is understood to belong to the specialty field of psychology. Rationality, like communication, involves exchange, involves a dialogical interaction in which some outcome results. Dialogical interaction presupposes energetic and substantive qualifications, mechanisms (transmitters and receivers), which together give rise by their reciprocation to new qualifications. Knowing that the energy of all substance is characterized in electrical potential, it seems logically evident that primitive energy was/is fundamentally electrical and disposed toward the creation of positive/negative polarities. The qualification of electrical polarization is a fundamental and primitive disposition that enables all dialog and communication, even as it is yet involved in all the physical/biochemical communications of complex vitality.

Primitively, bi-polarity is the root qualification-energy of *cosmic sexuality*. Cosmic sexuality is the primitive positive/negative attraction between the poles of electrical potential. The charge of polar potential is the dynamic of all communication, of all logical dialectic; as an axiom of quantum-electric cosmology, of paradigmatic nature, it may be observed that mono-polarity, for example, homosexuality, wants for vital, ambivalent potential. Patriarchy falls into the category of hyped mono-polarity.

A historical chronology of the evolution of rational complexity isn't attempted here, only a thesis concerning the fundamental dynamics of energy/matter by which changes in qualifications are amassed. History is itself a product of transformational necessity. Very first matter is minutely particulate and energetic. The word "quantum" here means every least particle, and the word "electric" means the energetic potential of all particles. When the term *quantum-electric* is used, it should be understood in this way.

Atomic/molecular assemblies are accomplished by particulate interactions that inhere and cohere the assemblies of atoms, namely, their nuclei (protons) and electrons. The process by which molecules are formed is a process in which atoms interact with each other because of energetic relationships qualified by electron-sharing. The qualifications of energy-states, which cause particulate attraction and repulsion, is an elemental kind of consciousness which opens to the communication of entities with each other because of electrical attraction and/or repulsion. By the process of communication and consciousness, new entities are amassed due to electrical potentials; and in the newly amassed construct, potentials for further interaction, new and different from those possessed in the original contributors, are amplified; by electromagnetic amplification, new signaling invites new connections and new openness. In this manner, active media of consciousness broadcast electrically new consciousness in waves of expanding potential.

Having a sense of the evolutionary taming of the explosive energy of the Big Bang gives us a sense of whence cosmic rationality originates and how it propagates. First energy was gross, untamed, massive, violent, unordered and purposeful only in its capacity to expand in the indeterminate infinity of an undefined cosmos. The capacity of open expansion endures as a defining characteristic of cosmic energy—spirituality—even in the highly reformed and refined subtleties of human consciousness.

Transformational necessity is cosmic reason's elemental purpose. This conclusion comes from experiential consciousness and from evidence of the evolutionary and unbroken continuity between first energy and contemporary conscious energy. The axiom of science states that the highly refined subtleties of Big Bang energy are derived qualifications already implicated potentially in original, unqualified energy.

From personal experience, we know that intended purposes of human consciousness bring about consequences intended and unintended. While it is true that humanly intended purpose may not correspond to cosmically ordered relationships governing energy utilization, it is true that human options remain open notwithstanding counter-intuitional purpose. Even though intentional human purpose may fly in the face if in-tensionally ordered nature, humans have the freedom to exercise free choices that may turn on them with unintended consequences. Under such circumstances, rationality may inform a change of direction and opt purposes that correct misdirection and realign human rationality so as to conform to the controlling hierarchies of natural order.

A continuity law of cosmic co-dependency may be defined as to the coincident necessity of quantum-electric inherency/coherency. This law testifies to the hierarchy of interdependencies and would say that: later evolved complexities, physical and psychical, e.g., sense organs, are expressions of prior established material/spiritual subsidiarities, *platforms,* that engage the prior energetic platforms of subsidiarities; except these are sequentially put in place in the genetic expression of a growing organism, subsequent complexities will fail. Natural hierarchies hold control over human hierarchies.

Signaling that comes from molecular complexities is electromagnetic; it is a wave complexity that is qualified by the molecular (quantum-electric) construct of the emitting agency. In the case of the "human molecule", the senses of the body, sight, hearing, smelling, tasting and feeling are all faculties assembled by and in specially characterized molecules that are selectively responsive to wave emanations coming from some other object, whether the emanation is a light wave, a sound wave, an aroma, a flavor or a physical contact. Signals coming from the objects are "signs" that are received in receptor organs. Signals may be invitations or they may be warnings, and based on experiences of outcomes, they will come to be identified with good and/or bad consequences.

Wave signaling is nature's universal manner of communication, of informing consciousness. "Grace" and evil are associated with signs and with the consequences of responses to the invitations and warnings of electromagnetic communication. Sunshine is inviting on a cold day, but it may be repulsive and physically damaging on a hot day. The sight of a fresh ripe apple is a sign that pleases the appetite, and when the apple is eaten, it promotes good health.

Sign (nature) and *grace* (nurture) are characteristic evidences of *sacrament*, which enable choice to distinguish and opt good rather than evil. Energetic inscriptions in the constructs of nature ("word" content, signaling) inform consciousness in relational consequence. This common phenomenon of information exchange is the reflective basis for the ritually celebrated awareness of religious sacrament. The presence of God in Sacrament is "cosmic presence", a sign of the grace of cosmic rationality. This quantum-electric fact is the basis for understanding the essential oneness of the *religious* and the *secular*; and, so long as human consciousness divides competitively the realms of the spiritual and the material it will continue to inflict on itself alienation, disconnection, schizophrenia and ecological degradation.

The Reason/Faith Relationship
(Reading the Codes of Quantum Relativity)

The cosmically script rationality of quantum relativity communicates intuitively in all of nature. Quantum-electric relativity embodies the "rosetta" code of spirit/matter. Consciousness is electrical sensitivity coded in matter. The superficial reading and interpretation of the cosmos easily obfuscate the natural codes of communication in self-aware consciousness. Superficial obfuscation gives rise to misinformed consciousness. In myths, for example, only informed glimpses of consciousness surface and not the whole of cosmic complexity. *Quantum theology* (belief/learning) is rationality in process. Faith premised in partial and misinformed consciousness produces misdirected religion, and actions born of partially informed faith/consciousness may become destructively harmful. Getting right the intuitive sense of consciousness in the Natural Order of Sacrament avoids self-injury and sets human relationships in conscionably right directions.

Mature faith *comes of age* with the maturity of adult rationality. Maturity carries with it the gradual, rational reconciliation of modern secularity's need for critical thinking and of Catholicism's demand for obedience to its authority. Faith and reason interpret consciously the codes of vital relativity. With maturity, faith and reason coalesce seamlessly in consciousness. The seamless rationality of age and experience is wisdom. Perhaps until the code of the quantum-electric "rosetta" script is cracked, the maturing of faith will continue to be stymied.

The sorry arrogance of institutional patriarchy in absolutizing the unfinished dialog of God with nature is theologically a sin of mortal proportion and historically of catastrophic calculation. Its warrant is nothing less than the attempted shutdown of intuitional rationality. Advancing together on parallel tracks, matter (prescriptive nature) and spirit (open consciousness) can and must acquiesce to mutual accommodation in the passenger coach of essential continuity. Prescriptive male exclusiveness must yield to intuitive female inclusiveness if the train is to avoid ever more consequential derailments. Getting civilization back on track may begin with a fair minded synthesis of luminary thought (wisdom), of great and near contemporaries, and with the bridging of the divisions between old, partially informed, and now unworkable world views, and fresher, better informed new world views.

The role of faith in religious life has always been somewhat enigmatic, and so has the role of reason. The conflict between them has been unsettling for centuries. In fact, in broad lines the Thirty Years Wars of Religions in Europe (1618-1648) might be characterized as wars between faith and reason. Two great proponents of reason and faith came in conflict with each other and played quite different roles in history. Both Desiderius Erasmus (1466-1536) and Martin Luther (1483-1546) were monks of the same Augustinian Religious Order as well as contemporaries, but of different national origin.

The personalities of the two were very much in contrast as were the reason/faith roles played out in their lives. Both were Renaissance luminaries at a time of high flowering of mercantile and societal growth. Erasmus particularly was a well-traveled person throughout Europe and his passion for humane tolerance and accommodation became his lasting contribution, among others of religious/historical importance. In the wake of Luther's passionate rebuke of ecclesiastical corruption the more quiet work of Erasmus is eclipsed. Erasmus chose to operate within the constraints of Catholicism even as he championed humanism and the cause of reason, whereas, as is famously known, Luther chose not to.

The imperial influence of the Holy Roman Empire everywhere held control over the minds and activities of the faithful; but, as the saying has it, *absolute power corrupts absolutely*. The turbulent after-effects that ensued from the war of absolutist faith with reason/humanism are not yet resolved. Institutional Catholicism still chafes against the claim of the faithful for a role in Church decision-making. Faith over reason played a dominant role within the Church and within the Kingdom for it was used as a religious and political tool to keep the people submissive to the theocratic State authority.

The famous first *father of the philosophy of history*, Nicolo Machiavelli (1468-1527) was an eager proponent of Church/State political control, and his personal ambition was to serve the Borgia Popes, and the De Medici Popes, to this end. His book "The Prince" is read yet for its strategies of political control.

The explosive mechanism that triggered the pent-up tensions of European society was the great and gathering weight of Church/State appetites for opulence, personal as well as institutional. Prelates and princes and the upper class of Church/State society were from the same families. Castles and cathedrals were joined in collecting taxes from the people to finance family, Church and State ventures. These ventures included the construction and elaborate furnishing of castles and cathedrals and the opulent living of the occupants, whose bureaucracies understandably flourished under the circumstances of absolute privilege. The Churches sale of promises (indulgences) of after-life rewards to the faithful for giving money to support Church-waged wars to promote the faith (the Crusades) was over-the-top. On October 31, 1517, Luther nailed his 95 faith propositions on the door of the Palast Church in Wittenberg, Germany, and he became a rallying polarity for the people against the monopolistic power of Roman Catholicism. The growing mood of Renaissance humanism championed the causes of reason against the imperial abuses of faith.

Luther is in a class of believers who belong to the "golden" age of faith, who, in his own way was even more single-minded, if that is possible, than Roman Catholicism. If he had a mean side it is understandable for the meanness he was confronting was older and more embedded in its arrogance than his; he affirmed the salvation principle of "faith alone", namely, that work does not contribute to justification before God.

Desiderius Erasmus was quite a different person who left a very different mark on history. He was one of many thoughtful persons of his time who came across tempered in their valuation of reason, with softer dialog and a more humane tolerance of differences. Theirs were voices of reason that meant to prevent the violent clash of religious/political ideologies. High on the list of dedicated humanists was Sir (Saint) Thomas More (1478-1534), Lord Chancellor to Henry VIII, King of England. Sir Thomas More was famous among other things for his faithful service to the king and to his defense of the faith, which accrued as well to the king's remembrance in history. Chancellor More is more famed for his opposition to the king's determination to divorce his Spanish Queen Catherine and the price he had to pay for it.

Charles A. Brady, in his historical novel, <u>Stage of Fools</u> (Chapter V, "Master Colet Preaches A Sermon, Morning, Good Friday, 1518", copyright 1953, E P Dutton and Company, Inc., New York), has a group of church leaders coming to Sir Thomas to ask him to write a thesis in defense of the Catholic faith for they knew he had the King's confidence. Sir Thomas and the bishops bandied back and forth over the primacy place of faith and reason. Bishop Fitzjames badmouthed Erasmus and drew a response form Sir Thomas: "He [Erasmus] defends our sovereign palisade of reason against the strong onslaughts of those who would take it by storm". To which Bishop Fitzjames objected, "Reason? Reason is a pagan goddess. We of the new dispensation have long preferred to speak of faith. There lies our pearl and our prize!" To which Sir Thomas rejoined, "I am afraid, my Lord Bishop, that Martin Luther would be in agreement with you on this point. What is it Martin Luther says? 'You must wring the neck of the beast, reason!'" More continued and made his point about the relationship of faith and reason: "If faith is Mary, reason is the Martha of human personality. They must dwell together as sisters, in amity and in the same household. Like man and maid betrothed, they must tread a married lover's dance, quarreling sometimes, but never severing from one another."

Sylvester L. Steffen

The Pattern is the Plan

A philosophical/cultural term much in vogue is *post-modernity*, which implies, whatever it is, that it succeeds *modernity*. The term "modernity" might generally include that period of the scientific enlightenment, which took upon itself the deconstruction of societal presumptions based on magic, myth, outmoded worldviews and theologies premised on them. In this "post" modern time, after centuries of acquisitions of new knowledge, after the deconstruction of age-old traditions of belief and practice, human consciousness is left confused, unsettled, even schizophrenic and uncertain of its belief in either religion or science. To many it seems that in our time chaos reigns. In science, an acentric cosmology has replaced the ancient Earth/Sun-centered cosmology, which long served religion and culture in its presumptions of centrism, staticism and absolutism. We are now told that in the universe there is no one static center of gravity, but that gravity has been inflated and expanded throughout the universe where its center is shared and its expansion continues. By analogy, religion no longer has a single static center of gravity, which means to say that religion does not reside in an elite, professional, theological focus group, insight or institution.

The centrist elitism of professions is not necessarily more authentic than the common sense of humanity; as a matter of fact, theology relies on the evidence of the "sensus communis" to authenticate the conclusions it formulates. What is the public to think? For, after all it has been the professionals who have constructed the prevailing worldview, and it has been the professionals who have deconstructed established cultural presumptions and brought us to the condition of dissolution we find ourselves in. Who will make sense of the chaotic consciousness we are heir to?

Human psychology has wrestled in the past and continues to wrestle with, for example, the psychic influence of human/star connections. Modern science has concluded that much of life's chemistry, including metals, originates in supernovae events, and that vital body substance is from stars. To the extent that materiality derives (is interactive with) from spirituality, so is consciousness. Obviously, there are faith implications to human/star connections for us today.

Back when astronomer/theologians were wrestling with Earth-centrism and Sun-centrism, they also wrestled with psychic connections (astrology). Today we are compelled to ask to what extent religion and science are psychically connected, for in the dualistic (Cartesian) presumptions dividing soul and substance, spirit and matter, etc., the religion/science (physics/metaphysics) split is still at an impasse and fuels wars between spiritual and material absolutists.

The Dominican Monk Giordano Bruno was burned at the stake (1600) because spiritual absolutists in Roman Catholicism insisted on the physical/spiritual centrism of cosmology and theology. Theologians and astronomers of the time were very sensitive to this worldview and the problems it posed in walking the line between science and theology. For example, Philipp Melanchthon (1497-1560), Tycho Brahe (1546-1601), and Johannes Kepler (1571-1630), like Bruno and their contemporaries, dealt with astrological and astronomical connections of faith and reason. (Ref: Kenneth Howell, "God's Two Books, Copernican Cosmology and Biblical Interpretation in Early Modern Science", 2002, University of Notre Dame Press, N.D., IN 46556). Beliefs and practices concerning astrological/astronomical connection raise the issue of *superstition*, which, among other matters, concerns extraterrestrial, spiritual agencies. In general, Roman Catholicism considers superstition to be a form of devil worship, a kind of idolatry and obsession with evil. Old dualistic obsessions in the matters of spiritual/material oppositions continue to stand in the way of the reconciliation of science and religion.

The matter of absolutism has to do with *fixations*, with the obsessive holding on to literalist beliefs rooted in misinformation and misstatements of reality. Erroneous faith obsessions are fixations called *fideism*. Fixations are beliefs that refuse to change even in the face of evidence exposing their falsity. Fideism is about fixation in spiritual and material matters, good and evil. Centrism, staticism, elitism and absolutism are examples of fixations associated with

21

religious belief and practice. Fixations, when found to be misinformed, may be damaging to faith and credibility. Institutional Roman Catholicism is adamant in affirming its belief that the institution (and the "infallible" pope) can make no mistake of judgment in matters of faith and morals. Given the evidence of history, there are those who would label this belief as an obvious fixation of cultured, institutional arrogance. While the condemnation and killing of Bruno isn't the biggest or only example of the Church's measured error in moral judgment, it is a compelling and obvious one, even if Bruno was more rudely arrogant than the Church.

The problem of the present time is, as Vatican II affirms, that a new, evolutionary, transformational worldview has replaced the long held static-centrist worldview, and that this new, conscious reality is with theological implications as to old errors and fixations. Human consciousness resides in the unitary mind/body venue, common to humanity, and spiritual/material connections are cosmic realities, a mystery from which humans arise and which defines the self's own mystery. *The Mystery of this Conscious Reality is the Cosmic Pattern. This Reality, because of its necessary priority in space/time change, anticipates the later subjectivities of its own expression and will always exceed their boundaries of understanding.*

In her book "MONA LISA'S MOUSTACHE, Making Sense of a Dissolving World" (Phanes Press, 2001, An Alexandria Book, P.O. Box 6114, Grand Rapids, MI 49516), Mary Settegast quotes what Dame Rebecca West considers to be the "dominant mood" of our time. In her view the mood of the time is a quest, "a desperate search for a pattern". (Pg 1)

Given the human penchant for dividing, dissecting, disconnecting and dissolving things, it is no wonder that human consciousness finds itself schizophrenically disjointed from the reality of its origins, and that it is in fact in a *desperate* mood. If the search is for a pattern, then it seems eminently sensible to go back in time and connect contemporary patterns of experience with patterns of beginnings, to the extent that they can be recaptured and understood.

In evolution, a new beginning is associated with endings. The ending of the age called "modernity" and its dissolutions provide beginnings and opportunities for post-modernity. "[W]hat we are referring to as the dissolution of form may actually be the disintegration of an outworn mode of human consciousness…the destruction of familiar realities and the multiplying of perspectives in our time is caused not by our position at the end or beginning of a cycle of time, but by the impact of an emergent mode of awareness…the integral structure of consciousness." (Ibid. Chapter 7, "The World is Evolving" pg 107). While few may deny that the world is evolving, more than a few might disagree with the observation that *modernity* is "…actually…the disintegration of an outworn mode of human consciousness". Many are so fixated in a habituated mode of human consciousness that they refuse to admit that it is "outworn". As a person of Piscean culture, openly anticipating Aquarian openness, I acknowledge that conscious fixations in staticism, centrism, elitism and absolutism are indeed "outworn modes", and that admission of their irrelevance can motivate cultural advance beyond their dead ends to the more open and fruitful consciousness of evolution and transformation. There is, however, an anciently recognized processing (Trinitarian), which works as a root process of transformation and openness, that might be characterized as a "cosmic pattern", and by which, the prosaic, the outworn, the absolute, the dead, can be resurrected from their pessimisms into a refreshed consciousness of renewing modes.

Human consciousness is at the threshold of *postmodern* insight that may in this dark hour lift consciousness and cosmic vitality out of its pessimisms. The resurrection of consciousness is enabled by the renewing patterns of cosmic transformation, patterns that are resonantly attenuated in harmonies of "trimorphism". Consciousness is an ever renewing "second enlightenment" of insight, which is energetically amplified in the resonance of communication and conscience; conscience is an ever renewing conviction, which is energetically amplified in the resonance of communication and consciousness; and communication is an ever renewing dialog, which is energetically amplified in the resonance of consciousness and conscience. Faith, hope and love are the resonant products and motivations of communication, consciousness and conscience, and except for their joined virtues, self-aware vitality cannot thrive. Trimorphic resonance is the Trinitarian *hypostasis* (understanding) of familial and communal harmony. These are the renewing bases of personal and social civility. Trimorphic resonance enables harmonic living, the unifying consilience of the secular and the religious collaboration.

Trimorphic resonance is no institutional or doctrinaire religion, nor is it a political expedient, rather, it is the processing pattern of nature/nurture, whose transformational advance assures and insures the mutual success (fulfillment) of personal and communal vitality.

Communal fecundity is pregnant with the seminal expectancy of new awareness, the egg-made-fertile in the dialog of opposites. The air of anticipation is charged with the electricity of Aquarian consciousness, not by backward looking batteries of centrist polarity, no matter how colorfully baroque their fireworks may have been in the past. The electricity, the excitement of the potential of new consciousness can make one tingle all over with the exhilarating feeling of transfiguration, the warmth and light of resurrection consciousness. By it faith is made ever more vital, more rooted, more expressive in its confidence. Quantum-electric insight is an innovating consciousness that accounts for the continuity consciousness in its space/time diversity.

Trinity Impact in Nature

Conscionable rationality is intuitive (divine) and dialectical (human), the refined, cosmic outcome of trial-and-error symbiosis. Religion, the spiritual energy of effective relationship, is Trinitarian, that is, a unitary spirituality that is reflexly and reflectively empowered. Theology in all its stripes may be thought of as the "philosophy of religion", which as an aspect of universal rationality relates to every human being in a universal way; however, the human/divine interactions of cosmic consciousness are qualified by particular (bioregional) circumstances. Thus, religion globally evolves under unique social and ecological circumstances that bear directly on the evolved consciousness of the particular bioregions. Biological diversification is characteristically local, yet this reality does not conflict with the "catholicity" of universal religion, of universal consciousness, and of particular connections under diversified circumstances. The Catholic doctrine on Trinity, while ultimately associated with *christic theology* and with bioregionally qualified *soteriology*, suits globally a universal consciousness, for the messianic, Christic role of Jesus in universal salvation links to his personal participation in divinity, the universal spirituality empowering the whole of cosmic rationality. Had Jesus lived in the Mexican bioregion, for example, he could have witnessed to the same Christology and soteriology, except, the incidentals of his witness would perhaps have been different, namely, characteristic of the Mexican bioregion. For example, at his Last Supper he might have indigenous drink, and instead of wheat bread he might have used corn chips.

The role of divinity in nature is—a *puzzlement*—to every involved observer; and of special interest to speculative Christian Theology is the special relationship of divinity and humanity in the person of Jesus—one person and two natures. Because of our Jesus' relationship, this matter pertains to us personally. This is of special interest because, if, as we believe, Jesus is the Son of God, then human beings have a firsthand view into the very nature of God. Jesus is a historically special case of the more general understanding that God is Self-expressed in creation, in nature.

In God's "natural" Self-revelation, cosmic rationality is not only not obviated by the divine economy, but rather, it is the vehicle by which the divine economy and God's Self-revelation are ongoing. Every encounter with God is, as Roger Haight observes, "experiential and historical". ("Jesus Symbol of God", 2000, pg 188, Orbis Books, Maryknoll, N.Y.) "Experiential" means personally relational, and "historical" means also *experience collectively recorded*, genetically/memetically, in the evolved codes of social awareness. In his landmark book, author Haight develops a rationality of salvation theology, which correlates with the thesis of quantum religion. The popularizing of quantum religion seeks more to effect bonding (attraction) than binding (compulsion), and, the valuation of all life-relationships instead of discriminatory devaluation.

Belief in Jesus' divinity is a source of information on how God relates to human beings for common well-being (salvation); on how Jesus' priestly anointing, messiahship, exemplifies the way, the truth, and the light for everyone; and how interactive divinity and humanity work in the God/man relationship. The professional term for *salvation theology* is "soteriology"; the theological term for the universal (priesthood) vocation of baptism, the Christic vocation *to serve,* is "Christology"; and the term for the union of the divine and human nature(s) of Jesus-person is the "Hypostatic Union". All the above pertain to every person insofar as every person is a "child of God" even as Jesus asserts everyone to be.

In the insightful theology of Teilhard de Chardin, the essence of Christology is conscious growth of humankind into a new "phylum of love". Personal and communal self-giving in the word/work of human welfare, as it is situated in the evolutionary ecology of life, is the essence of salvation as communicated by the "cosmic" Christ. Such self-giving is a commitment of love, whose ultimate fullness (Pleroma) is achievement of the full potential of the self. The fulfillment of personal *pleroma* advances an authentic sense of harmonic living and creates an atmosphere (noosphere) which is charged with exemplary energy, and which radiates positive motivation.

It is also clear to the involved observer that life on Earth is structured in an *ordered* way. "Hierarchical" arrangements are prioritized in order of occurrence and importance to well-being. From personal experience we know that we are individually *ordered*, but also *free*. The purposeful ordering of interdependencies can be characterized as inherent patterns of perfectibility that are open, yet inwardly driven by intensional forces, e.g., genetic, of transformational continuity. Purposeful evolution is the subtle insight of intuitional continuity, the rising consciousness of energy and matter, of soul and substance in process of perfecting. In the sense of intuitional continuity, *christic anointing*, the rationality of cosmic purpose *subsists* under/in the process of human/divine perfecting. Jesus reconciles in his person the divine and the human, the intuitive and the dialogical.

The Trinitarian consciousness of divine personification, as an Agency within and beyond nature, is an early theological awareness of Christianity, as evidenced in the final instruction of Jesus to his Apostles was, "Go therefore and make disciples of all nations, baptizing them in the name of the Father and of the Son and of the Holy Spirit". (Matt: 28-19).

In the hierarchical order of nature, every genetic adaptation is not of equal consequence. Some are more *substantial* than others, in the very literal meaning of the word "substantial". Some genetic changes serve as platforms that qualify all subsequent developments; case in point is the brief window of opportunity available for laying the foundations of language skills in the early formation of a child; if this window of learning is missed, the child's language skills may be compromised for life. In the case of the evolution of land animals, the ability to breathe and use oxygen (mitochondrial activity passed on by gene codes of maternal origin) in the controlled combustion of food and the recovery of food energy obviously ranks high in the hierarchy of the genetic ordering and perfecting of life. Except for prior hierarchy in genetic patterning there would be no land animals.

Traditional Catholic Theology holds that Jesus Christ is not just "of similar substance" with the Father and with humans, but that he is of "the same substance" [Charles Dickinson, "THE DIALECTICAL DEVELOPMENT OF DOCTRINE A Methodological Proposal", 1999, printed by Pryor Pettengill, Ann Arbor, MI 48107, pg. 199]. In the person of Jesus, the divine nature and human nature subsist "unconfusedly and unchangeably", as well as, "indivisibly and inseparably"; this characterizes the Catholic Doctrine of the Hypostatic Union of divinity and humanity in the person of Jesus.

The concept of *hypostatic union* is formulated in such a way that it is consistent with certain other theological principles, including the monotheistic belief of one God. Other principles include: the fact that Jesus lived a *fully human life* and that Jesus simultaneously possessed *divine perfection*. On the basis of this latter principle, it has been a tendency of Christian tradition to assert the highest conceivable level of divine attribution working in the person of Jesus; reconciling his full humanity and his full divinity, together working simultaneously, is more than mind boggling, it is at the heart of human/divine relationship and the *subsistence* of two natures in the one person of Jesus. It has been a historical challenge to put in satisfactory words the understanding of how Jesus' human will *freely* operates simultaneously with his fully operable divine will. If it is believed that Jesus' divine will dominated the judgment of his human will, then, it may be argued that Jesus wasn't really *fully human*.

The reconciling of the human will with the divine in the human purpose of self-perfecting is a matter that cannot be disconnected from cosmic purpose. How do the human (natural) and the divine (*supernatural*) relate to each other in the transformational cosmos? How do human knowledge and divine knowledge function at the same time in a human and divine manner in human beings? Intuitive sense tells us that it must. How it does is profoundly beyond full human understanding.

If the transformational universe in its self-perfecting is script by the nurture of divine purpose, what does the manifestation of divine purpose in nature tell us up until now? The consciousness of experience, a spiritual complexity, is substantively embedded in the transformational continuum; it tells us that everything composing the universe, including everything living, is, in process and in structure, both "ordered" and "free". *Ordering* tells us that we wholly belong to and depend upon Earth and the universe. The intuitional insight of experience confirms that Earthlife relationship is ordered *communally*, that is, subject to the exercise of free judgment in the details of interpersonal relationships. Because Jesus was fully human, he made personal, human choices as to his interpersonal relationships. In regard to these, Jesus' reflective human judgment measured his options and his choices in terms of *well-being*. Choices of well-being reflect the intentional purposes of universal consciousness in which divine purpose is embedded, inscripted. Scripture narrates a hierarchical experience in Jesus' life where human considerations were subordinate to the divine, and where Jesus showed for all times an example of human will tuned in its decision-making to the divine. We can discern from Jesus' living that every person is competent in his/her *baptismal* virtue to exercise personal conscience from the perspective of divine will. The *temptation* of Jesus before entering his public life exemplifies the critical interaction of his human will with his divine. We can put ourselves in his circumstance and realize that we too have the same freedom and the same *ordering* that enables us to respond to life's choices even as Jesus did.

The scriptural account of Jesus' temptation is put at the beginning of his public life. The Gospel story has the devil taking Jesus to a high place where he could see the world laid out before him as well as the options open to him in terms of his interaction with the world. Jesus was born into this world submissive to the same basic genetics that control in every human life. This is the "anointing" of inheritance, which controlled in Jesus' life, even as it controls in every human life. Divine intention is discernible to human knowledge through the ages in the Word as it comes to everyone in the implications of genetic revelation.

Jesus' temptation sets the stage for his work by challenging human/divine will. All the subsequent events of Jesus' life result from his life-choice at the outset. As we know, the choice of Jesus was to self-identify with the downtrodden, the dispossessed, the sick and suffering. *Sensitivity* was his passion—his motivation for speaking out against empire/temple sponsored oppression. Jesus' *preferential option for the poor* was a human/divine choice; a choice, which is truly human and truly divine for it confronts the empire and the temple. His harshest words were consistently leveled against the pretenders of religion—more so than against the pretenders of political authority. Politically, the Roman Empire ruled with a heavy hand, and religiously, the temple imposed heavy expectations of ritual and detail that dominated every aspect of peoples' lives from birth to death.

There are aspects to the account of Jesus' *temptation* that apply today as they applied in Jesus' time. The harsh domination of government and church continues to oppress. The violence of theocratic dominion, associated with religious orthodoxy, can be as vicious and destructive today as it has ever been in the past.

Jesus had imposed on himself a 40 days fast and retreat into the desert, which immediately preceded (or accompanied) his temptations. His self-imposed fast and retreat may be understood as a kind of *right of passage* by which Jesus puts himself to the test in preparation for his life's work. It was a sort of check on the maturity he possessed, whether or not he was ready to face life's ordeals. The Gospel of Mark does little more than make note of the fact of Jesus' self-imposed ordeal. His comment on it is to the effect that Jesus passed the test successfully: "He (Jesus) was with the wild beasts and the angels waited on him"—a way of saying that Jesus confronted his demons with divine insight. (Mark I: 13).

Jesus' temptation, as recorded by Matthew and Luke, may be seen as having four components. 1.) Jesus passed the starvation test: he was so hungry he envisioned stones turning to bread, but he brushed the illusion away by saying "not by bread alone does one live but by God's word". *Jesus affirmed the priority of the spiritual over the material.* 2.) The second component deals with control over the natural resourcefulness of Earth. The need of people for food creates a ready market for the diverse fruits of the land. In Jesus' temptation he is assured that control over these could be his if he committed himself to interests of profit and control over Earth-produced commodities. Nature's largesse is a gift of natural/divine providence and is intended to serve common need. Putting the will of self-advantage ahead of the common good is an imposition of human will on the divine. *It is wrong to substitute self in the place of divinity.* "God alone shall you adore; him alone shall you serve." 3.) The third component involves the consideration of temple service. From the faith perspective of temple service Jesus considered the favored role of the rabbi whose position of temple dedication commanded the respect of the people. In parading divine wisdom before the people he could enjoy the praise of the people and the assurance of being well provided. To this prospect Jesus answered by quoting an older scripture, which admonished that it is wrong "to put God to the test". Arrogating one's own person over other persons is an oppressive act. Jesus would not add to the burden of self-arrogation on the people, already excessively burdened by oppressions of Empire and Temple. 4.) In the fourth component, scripture tells that Jesus came to personal clarity. His certitude of mind closed the door on the devil, "who left him" expecting that there may be a future opportunity. With this clarity it became a certainty for Jesus what he would not do in his lifetime; what remained was for him to clarify what he would do for his lifetime. By his example, and in his later words, "What you do to the least of your people, you do unto me", Jesus revealed his life's option, namely, to join the company of the least, the rejected, the suffering, the poor, the hungry and the sick. Jesus' preferred option, which he expects to be the option of his followers, is to pursue the *kingdom within*, the role of social, spiritual betterment, which alone has the power to enable people in personal self-fulfillment. Joining the worldly kingdom-builders only aggravates public oppression. Spiritual clarity is the basis of effective action.

Wisdom is the practiced facility of being able to prioritize understandings of relationships—putting things in their rightful place of importance. Wisdom is a sense for right order and a facility for engaging knowledge, the understanding of things, in a way that sustains and advances right order. Good sense, reliability, even temper, tranquility, love for others, are signs of a life committed to the pursuit of wisdom. The wise person lives free of fear and by his/her own truth that radiates the light of inner authenticity. Such radiance is the radiance of divinity, which has the power to transfigure everything. It identifies *Christic authenticity.*

A person of wisdom, of altruism, harbors no ill thought toward anyone, rather she acts with sensitivity toward everyone. Sensitivity is a seed of good will that roots in the hearts and actions of others. Sensitivity, good heartedness toward others, is for all the beginning of wisdom, the grace of Godlikeness. Parents especially need to act with sensitivity toward each other and toward children so that the grace of Godlikeness becomes the experience and habit of children for their lifetimes. It is well for parents and teachers to heed "not to give children stones for bread, and...not to bore children".

We are born in ambivalence and characterized in ambivalence. But under the spiritual motivation of nurture, positive and negative, our ambiguous proclivities are fleshed out by personal choosing and are exercised and activated. Where the *proclivities of darkness* prevail on personal will, personality becomes disposed toward darkness, negative conduct; but where the energies of *light* prevail in the nurture of judgment, proclivities toward light, toward positive doing prevails. Parents, educators and peers have the power to push unformed minds and communities either in the direction of light or darkness. Motivation in life is greatly a matter *of nurture* for human personality is born responsive to the word and example of others, for good or for ill. It is left up to each individual personally to affirm light or darkness; light invites goodness, darkness invites evil. In so many ways *we become what we experience.*

Personalities become conflicted when cultured self-motivation clashes with the altruistic message of religion. A personal lifetime is occupied with the labor of distinguishing darkness from light. Tragically, people committed to darkness, to societal destruction, under pretenses of religion, become agents of evil, intent on corrupting the minds of the young—especially under circumstances where great social inequities prevail. These times witness to such societal catastrophes.

Order and freedom are codependents; one depends on the other. Good and evil are consequences of choice; order is a consequence of choice and so is havoc. By knowing and anticipating the causes of order and chaos we become equipped to make choices that secure and promote right order. Health, for example, involves practices that sustain well-being. If we are ignorant of health choices, and/or disregard the consequences of willful actions harmful to health, like smoking and doing drugs, for example, we are administering "evil" to our bodies, the consequence of which is to damage their healthful order. By making bad choices we can bring diseases of mind and body on ourselves.

We are free to make or not make choices. We are free to remain ignorant of the consequences of choices; we are free to inform ourselves of the consequences of choices and we are free to opt choices having good consequences or choices with damaging consequences, whether to our own bodies or whether to Earth's living networks. Freedom is an agency of order but also of havoc. The saying has it that *the devil is in the details*. If we habitually do things without regard for the details we *set the devil free*.

If we let altruism motivate us we will seriously seek to know the consequences of choices we make. This is the way of wisdom. As to wisdom and understanding: if we ask, we shall receive; if we seek, we shall find; if we knock, it shall be opened to us. Divinity is living by wisdom and understanding; the opposite of divinity, the diabolical, is the avoidance of wisdom and understanding and the careless trashing of essential order. With the exercise of wisdom and understanding comes the *rise of consciousness*; in the neglect of wisdom and understanding consciousness and conscience are deadened—hope and love are undermined. True religion is not exercised except in the conscious quest of wisdom and understanding, the guardian of order and freedom, the grounding of conscience and love. Religion isn't an institutional thing, it is a personal/social thing, the intension/intention of personal relationship.

I trouble my consciousness over the realization that we are ordered and free; it is my sense that we all should trouble ourselves about the relationship of freedom and order. How are we ordered? How are we free? We are ordered from within and from without. Our without is tethered to the skeleton within. But the within is more than a skeletal infrastructure. The order from within is in-tensional, is particulate inherency that is ordered by its own intension, which draws attention to the intention of the infrastructure. A seamless whole, intension, attention and intention, (communication, consciousness and conscience), are spiritual agencies advancing natural edifications.

For all our preoccupation with matter, all reduces ultimately to spirituality. And so it is with "resurrection". Subsequent individuality is but a *resurrection* from the dust of prior individuality, which has surrendered its photons to the rainbow. That is why there is hope in resurrection. That is why the rainbow represents hope—because it is the evidence of life reborn from death. Our *particularity,* the particulate construction of our body molecule, is ordered by laws governing the constitution of life by light; particulate construction is liberated by death and the bonds that hold light and dust together are released. Death lets light escape from dust, and lets dust to return whence it came.

An understanding of this truth allows us to say "O blessed death". Except for the release of light, there is no resurrection. In the realization of death's granting to us release, escape from the tethers of order, we return to the realms of our origins. The order of particulate tension, the tension between ordered consciousness and open consciousness, is the physical state to which we are bound until death liberates to a fully psychical state, which is too open and free to be comprehended in the strictured spaces of tensioned consciousness. Death is but an open door, a transitioning from an ordered psychic realm to a fully liberated realm of infinite consciousness. Words fail describing destiny for destiny belongs to infinite experience. Experience tethered to constricted order pales before experience whose order is unstructured. The ultimate liberator of constricted physical order is death, whose only agendum is spiritual freedom. *Where is death's sting?*

I have come to believe that we are well advised to live our lives in recognition that our afterlife should receive priority consideration. Whether aware of it or not, we live more for future life than we do for this life, if for no other reason than that afterlife exceeds infinitely the time of our presence on Earth. If I really believe in resurrection I have to believe that my afterlife is of more lasting importance than my present life. This conviction is a powerful motive to live seriously and with purpose so as to make the most of resurrection values in future life. No personal life should be a drag on rising consciousness. Serious and purposeful living is perhaps less from the perspective of winning personal reward in future life than from the perspective of "making way" for "Second Comings" of "other Christs" who will replace me when I die. If we understand and believe what Jesus taught us, we know that every newborn is in God's mind a Christ of

Second Coming. We should see ourselves in the role of John the Baptist, "preparing the way of the Lord".

So I use my years in the expectation of the resurrection of consciousness. Surely, the same motive drove Jesus to leave a legacy of liberation theology. Reason, faith and purpose edify resurrection theology in which every newborn is recognized as a Christ of Second Coming. The kingdom of God is within. Personal consciousness like a sprouting seed has the power to bring the kingdom to fullness. It inspires us to do to others what we want others to do to us.

This understanding came to me early in life and instilled in me the desire and intention to be a priest so as to serve others in their quest of self-discovery and authentic living. While I ultimately decided against becoming a priest of the institutional Catholic Church, I believe that I have been faithful to the original calling of grace within and to the vocation of helping others discover the "kingdom within" and what faithful religious living means. If my message is true and if it endures in others, it will be not just my personal resurrection but it will be a revitalized resurrection in others.

The Vatican Councils & Cosmic Rationality

Cosmic rationality is the essential continuity of universal consciousness, which traces back to the unqualified potential of Big Bang energy. Cosmic rationality is the spiritual dimension of highly qualified quantum relativity. Contemporary religious consciousness is an evolved refinement of cosmic rationality. Except modern rationality is faithfully grounded in an authentic consciousness of its true origins, it can stray from the rationality of its evolved purpose. Misinformed worldviews can produce inauthentic and destructive expressions—as our times attest.

The trademark word of Vatican II is "aggiornamento"—*updating*. If a trademark can be attributed to Vatican I it might be "resistamento"—*fixation in authoritarianism*. The mindsets of the two Vatican Councils could hardly be more contrasting. The pope of Vatican I was Pius IX, Napoleonic in personality and monarchical in determination. The pope of Vatican II was John XXIII, democratic in disposition and collegial in his vision of shared authority. The spiritual/political contrast of these two popes and the works of their two Councils are still having Church-shaking repercussions. The oppositional nature of their Council differences might be captured in a quote from the pope of each; Pope Pius IX said, "I am the Church, I am Tradition", (John W. O'Malley, S.J., "The Beatification of Pope Pius IX", AMERICA, A Jesuit Magazine, 8/26/00-9/2/00, Vol. 183, No. 5, Whole No. 4496, pp. 6-11, America Press, 106 W. 56th St., New York, NY 10019-3803), his affirmation to the Council with regard to his vision and claim for papal infallibility; Pope John XXIII says, "The human race has passed from a rather static concept of reality to a more dynamic, evolutionary one. In consequence, there has arisen a new series of problems, a series as important as can be, calling for new efforts of analysis and synthesis." Gaudium et Spes, Introductory statement, No 5, paragraph 4.

Among other things, the First Vatican Council is notable in its affirmation of clericalism, that is, in affirming the male exclusiveness of the institutional hierarchy in dispensing divine grace through the rituals of the Sacraments. In contrast, the Second Vatican Council affirmed the doctrine of the "People Church" and shared authority in grace possessed in the whole membership. The lines of distinction, between the people vs. the hierarchical Church and of shared (distributive) authority vs. exclusive (monarchical) authority, could hardly be drawn more sharply.

The cumulative ferment that set the two Councils off in different directions is generally associated with acculturated fixations surrounding notions of materiality and spirituality, reason and faith, and science and religion, in short, over the spirit/matter dualism and the "Cartesian" divide that evolved since the challenge of Copernicus, Bruno and Galileo on faith fixated in a static, Earth-centered cosmology. The fixated theology of authoritarianism (Vatican I) chose to hold firm to its cosmology/philosophy (Aristotelian) of staticism and centrism. To the extent that Church refuses to move away from its ancient fixations, it continues to lose credibility in the public mind. That the fixations of the Church of the First Vatican Council have become irrelevant to modern consciousness, notwithstanding authoritarian stirrings to the contrary, seems to be ever more obvious. Even though Roman Catholicism positions itself as a dam against the tides of changed thinking, change will not be stopped. The ominous threats of Mounts Etna and Vesuvius are powerful metaphors of the state of Roman Catholicism, which, like the boot of Italy, floats on a hot bed of restless lava. Containing the tides of change can be sustained only for so long; eventually, the blockages will be breached, and perhaps with catastrophic consequences, unless accommodations for change have been made.

Quantum science may provide insights to a new theology that allows for "up-dating" and mitigating the cumulative pressures of needed change that has been postponed. Elementally, the cumulative pressures of quantum reality are also building in Etna and Vesuvius, and in human consciousness, to the point of eventual eruption. Knowledge and consciousness are no more containable than the pressurized magma upon which Italy floats. The Church will do well to see the parallel and to seek a course of evacuation from its futile fixations before eruptions cause cataclysmic consequences.

We live in a quantum-electric universe. The "spiritual" dynamic of every least and great cosmic structure is quantum-electric, electromagnetic. Consciousness it self, is a highly perfected subtlety of the quantum-electric dynamic. Until we recognize, and intentionally cultivate, our essential relativity—connectivity to the whole cosmos—we will continue to arrogate wrong notions of self and God, and destroy and devalue ourselves and the Earthly/earthy definitions of our personal origins, dependencies, and destiny.

Institutions, no less than individuals, must, by intentional process of rationality (communication/consciousness/conscience), engage conscience in an updated manner. Fixations in old presumptions, that no longer inform, fail to enable conscience to deal with the necessities of contemporary realities, and in effect misdirect people in their essential connections. Faith, in the exercise of misguided fixations, is "fideism", which becomes a power of self-deceit and destruction. *Justification by faith* is a rationality solidly rooted and engaged in the consciousness (in-tensional and intentional) of the quantum-electric cosmos.

Faith isn't a fixed absolute. Nor is justification. Together they constitute spiritual certitude that continues to affirm the conscionable place of agents in the conscious complex of the transformational cosmos. By the intentional engagement of rationality, humans attain and experience standing in cosmic/divine purpose. Faith is experientially rooted in the evolutionary history of the cosmic complex. The conscious spirituality of the process of rationality—faith/hope/love—is an energetic continuity of the evolving energy/matter complex, a continuity that essentially links the transformational universe back to its origins in and connection to Big Bang expansion. All well-being is sustained in transformational, conscionable relationship. As in the beginning, so now, conscious cosmic well-being, as it pertains to Earth life, originates in and is sustained by the "conscionable" relationships built upon evolving spirituality, faith, hope and love.

Vatican II intentionally set out to dislodge Church from "fideism". This expressed intention was the Second Vatican Council's way of relating to "modernism", which to Vatican I was plain and simply a diabolic misdirection of modern culture and science. Fixations are best resolved by rational dialogue, which separates the wheat from the chaff and preserves truth's viable grains, for, in fidelity to cosmic rationality, "middletree" vitality is secured.

Pursuing Truth

Most people don't much concern themselves with the minutiae of philosophy and theology; much less do they trouble themselves to think about how philosophy and theology interface each other. A problem with professional disciplines, in general, is that they address fields of specialized interest and develop in-depth approaches and procedures that to the ordinary person may seem beyond the realm of common sense and practicality. And so the turf claimed by professionals goes largely unchallenged by the lay outsider and is left exposed to insider biases. Insiders often become fixated in the profession's evolved logic and its established presumptions. Though the observation may come as a shock, the fact of the matter is that the evolutionary dynamic of professionals tends to be incestuous for their "truth" is kept within the *family*. Some might argue that patriarchal exclusionism in the Roman Church and papal bias in the appointment of bishops and cardinals create an incestuous mindset of male myopia, which compromises the Church's vision of truth.

If everyday people are meant to live by truth, and if philosophy and theology are professions that purport to specialize in the knowledge of truth, then the objective of truth takes precedence over professional fixations made on established turf, for these fixations may misinform and mislead the profession from truth's path causing the profession to be an agent of public misdirection.

In their understanding of the place of humans in the cosmos and how the cosmos works, whether in the scale of the macrocosmic or microcosmic, human theologies have a lot of catching up to do. Theological minds of Christian traditions are still fixed in an ancient worldview, while public awareness of realities have changed and moved on. Present day movement within theological circles seems to suggest that a quantum leap is on the verge of happening. Well almost. Theological consciousness is only now allowing itself to move from one center of consciousness, Earth, to another center, the Sun. Theological consciousness hasn't yet adjusted to the Copernican/Galileoan heliocentric model, while science has already moved on to the acentric understanding that there are many galaxies spiraling in ever expanding rotation. It seems that Christian theologians for the most part don't realize that they are occupying themselves with a new but outdated centrism. What macrocosmic and microcosmic insight is revealing is that the cosmic "center is everywhere and its circumference is nowhere". In the inflationary expansion of galaxies from within, new centers are constantly coming into existence, like dividing cells, which divide the center they seem to have and share it with the newly formed cell. In the ongoing processes of center sharing new continuity worlds are constantly created.

This new cosmological insight is actually prior to, or at least contemporary with, the Copernican heliocentric understanding. The Dominican Monk, Giordano Bruno (1548-1600), arrived at the cosmological insight of *homogeneity* (the properties of matter are everywhere identical) and *isotropy* (that the cosmos appears the same from any point of observation) as "the consequence of his discovery of cosmic acentricity and infinity". [Ramon G. Mendoza, "The Acentric Labyrinth", 1995, Element Books, Inc., P.O. Box 830, Rockport, MA 01966, pg. 74]. "Bruno's discovery of the infinitude, isotropy, and homogeneity of the universe...has been carried by Linde to its ultimate consequence. The All is no longer necessarily a sea of billions of galaxies and clusters of galaxies; the All may be an infinite ocean of infinite universes!" (Ibid, pg, 184). This sounds like St. John Damascene's definition of God, "a sea of infinite substance". Because of the overlap of Bruno's radical cosmology into the centrist politics of theology, the Roman Inquisition, under Robert Cardinal Bellarmine, SJ, felt compelled to condemn the visionary monk and burn him at the stake in 1600. This event had a chastening impact on dissenters, then and now, "...[T]he autonomy and independence of reason from religious supervision...was the decisive reason for Bruno's condemnation and execution, since his position posed the most dangerous threat to the power of ecclesiastical authorities should they ever lose their tight grip on scientific inquiry." (Ibid, pg, 167). Who lives by fear lives by another's truth; timorous religious writers fall into this class. Until now, Catholic writers who write on the new cosmology, e.g., Cletus Wessels, OP, "The Holy Web, Church and the New Universe Story", 2000, Orbis Books), AMERICA Associate Editor David Toolan, SJ ("At Home in the Cosmos", 2001, Orbis Books), and the historian Kenneth J. Howell's "God's Two Books, Copernican Cosmology and Biblical Interpretation in Early Modern Science", 2002, University of Notre Dame Press, Notre Dame, IN 46556), leave out any mention of Bruno for what might seem obvious reasons, namely, that the Catholic Church isn't ready to admit to any wrongdoing in the Bruno affair, and that Catholic "orthodox" writers are fearful to breath the word "Bruno" even if it means misrepresenting history and denying credit where credit belongs. On its face, history documents a conspiracy of silence by Church in its condemnation and discrediting of Bruno and its refusal to confess any wrongdoing for this specific injustice. The matter is complicated by the fact that before his defeat at Waterloo

Napoleon had plundered the archives of the Holy Office and made off with the records of the original acts surrounding the Roman trial of Bruno; after Waterloo, the pope reclaimed them, and on the way of their return to Rome from Paris the records were lost. (Ramon G. Mendoza, Ibid. pg. 52).

In light of understandings of *homogeneity,* new understandings may surface as to quantum relative meanings of "transubstantiation", for example. Homogeneity says that cosmic stuff is the same throughout the universe. Subatomic/atomic particles dynamically constitute and sustain all substance. The great and apparent diversity of energy/matter, manifested also in Earth life, results from the substantive and subjective joining of energy/matter "complexities" (molecules). Thus, in the communication and linkage of energy/matter complexities—*transubstantiation*—the formation of new molecules continually processes.

The principle of homogeneity states that the human body is composed of the same substances as are stones, soil, plants, grains, *bread*. The reality of homogeneity makes it possible for us to say "really" as Jesus said over bread and wine at his Last Supper, "This is my body. This is my blood." It is by the natural mystery of transubstantiation that our bodies' substances (e.g., components of DNA) are "in" bread and wine and that bread and wine "become" our bodies. The natural *sacrament* of cosmic word/work affects the reality of all transformation. In this cosmic "sacrament of God Present" all vitality shares, in special ways, degrees of God-consciousness and God-likeness.

The communication of prior substances "works" the consequence of bringing about new creations of soul/substance. In the *eucharistic* working of natural sacrament (transubstantiation) all life experiences a perpetual making-over. Every atom in the human body is systemically renewed over time so that, by reason of the food we eat and assimilate and by reason of the air we breathe and burn, every particle in our bodies comes to be replaced. We are constantly being reborn into new creatures, one particulate component at a time. This realization of substance regeneration should motivate in us an *intention* of spiritual renewal that corresponds with our material (intensional) freshening.

41

The principle of *isotropy* states that we, individually and personally, can truthfully say, "I am at the center of the universe", even as any other living, self-conscious being in any other galaxy, on any other planet, can say with the same certainty. And each of us, gazing into the endless skies will see and experience essentially the same view, namely, that in all directions beyond us are countless swirling galaxies in continuing states of birth, expansion and dying— *a sea of infinite substance.*

It shouldn't be surprising that philosophers and theologians argue about truth and about the primacy and discipline the one tries to put on the other. Philosophers tend to argue for the *primacy of reason* while theologians tend to argue for the *primacy of faith.* And so, the battle drawn between science and religion reduces to the truth-claim of reason over faith, and of faith over reason. Perhaps the truth lies in the middle, namely, that faith and reason are both needed and that truth's claim is validated only in the harmony of their mutual insights. The mutual engagement of religion and science in the quest of truth is a dance of Faith and Reason; the insights of both are needed, and the breaking off of the engagement by either party compromises truth. Reason and faith are both special kinds of consciousness; but neither is a consciousness independent of the other. Our times, perhaps more than any other, are psychologically unsettled by the diseases of negative culturing (reason vs. faith), which aggravate disease proclivities, for example, in matters of sexual ambivalence. The sustaining sensibility of conscious vitality's ground state (femininity) has the potential of mitigating alienation and its attendant provocations.

St. Thomas Aquinas advanced in his theological summa (drawing heavily on Aristotle) arguments of philosophy to prove, for example, the existence of God. To a philosopher it is illogical to use the faith presumptions of theology as premises in philosophical argumentation. "Faith on faith" is doing theology, not philosophy. [Charles Dickinson, "The Dialectical Development of Doctrine, A Methodological Proposal", 1999, pg. 108, printed by Pryor Pettingill, Inc., Ann Arbor, MI 48107]. By so doing Thomas Aquinas is accused of forfeiting his claim as a philosopher. Traditional (Scholastic) philosophy (theology), like the "Cynic" philosophy (quoting Cardinal

Ratzinger, Ibid Pg., 111) upon which it is premised, obsesses on death, whereas, Vatican II theology in contrast focuses on issues of life and asks life questions. "What is man? What is the meaning and purpose of life? What is upright behavior, and what is sinful? Where does suffering originate, and what does it serve? How can genuine happiness be found? What happens at death? What is judgment? What reward follows death? And finally, what is the ultimate mystery, beyond human explanation, which embraces our entire existence, from which we take our origin and towards which we tend?" (Charles Dickinson, Ibid. Pg.111).

The pursuit of truthful (informed) answers to the above questions belongs to both philosophy (science) and theology (religion). The purview of both disciplines obviously overlaps, and in the overlap they necessarily engage each other. Truth, the area of overlap, belongs to common consciousness, the ultimate venue also of philosophy and theology. Hard (scientific) information is healthful for it may confirm and/or correct the misdirection of cultured traditions.

We are born into *hypostatic* (divine/human) relationships that are based more on intuitional wisdom (faith) than on reflective understandings of personal intelligence. Evolved hypostatic consciousness is a subtlety of cumulative experience spiritually embodied in the holistic complexity of cosmic consciousness. Hypostatic wisdom is an important aspect of cosmic rationality, which functions at intuitional depths more profoundly than reflective consciousness is capable of apprehending. While water in a cup participates in the infinity of the sea, the cup cannot contain the sea. The ultimate outcome of divine/human connection exceeds experiential comprehension.

The Sea of Wisdom holding cosmic insight and purpose is the God of all being that anticipates and implicates all that ever was and all that ever will be. Properly awed by the face of divinity, what humans best can do is to remember, celebrate, honor and love, content in the knowledge that the awesome grandeur of all being/becoming is the cause of our own origin, experience and destiny.

In the harmony of conscionable living we intentionally enter into the resonant anticipation of hypostatic union, and by so doing, we attenuate and amplify natural/divine perfections perpetually rising, not merely in individuals but collectively in humankind. The harmony of life on Earth is "heavenly" participation in divine relationships. We can revel in the unending harmony of the baroque Choir of the All and thereby contribute our bit to perfectible reality.

Is religion necessary? Yes, because religion is a natural aspect of conscious, communal self-expression. It is an intuitive need of experiential consciousness. It is an expression of the purposeful intention of cosmic rationality. It is the intension of work needing to be expressed in word; it is word expressing itself in work. As the expression of conscious authenticity, religion is both cause and effect of rationality, of conscience. Religion is an evolved and evolving consequence of authentic self-expression; as such, it is a consciousness of cosmic rationality, an interior spiritual compulsion needing exterior expression.

Cosmic rationality serves the purpose of well-being. Rationality has various cause-and-effect aspects. Religion (concern for self and other), education and public service are all aspects of expressive reason/faith consciousness. Taken together, the exercise of these makes up *worship*—doing for others what has been done for us. Religion is the exercise of communal celebration, the celebration of understanding, thankfulness and of giving credit. Religion *recognizes* that personal existence is an absolute gratuity of others' doing; it *gives thanks* to God and to all who have preceded us and made us who we are; and religion *responds reciprocally* in service so others benefit as we have benefited. Remembrance, recognition and thanks are done in communal celebrations of ritual and song. These are the substance of public worship.

The enabling of others involves the giving of self in service to others in many ways: being teachers even as we have been taught; being providers even as it has been provided to us; bringing healing even as we have been healed; giving joyfulness even as we have received joyfulness; consoling even as we have been consoled. Personal expression of all of these is necessary for self-fulfillment and community building. Sensibility and insight tell us it is so.

In living religiously we mirror Godlikeness. The works of good living express religion, the soulful amplification of expressed virtue (*virtus*, spirituality focused on well-being). Neither institution nor any collection of people has exclusive claim on God or truth; authenticity in relationships obliges all equally. Shared spiritual authenticity is leaven that gives rise to Godlikeness.

The *religious* community is composed of likeminded people harmonized in right relationships and spiritually empowered to the purpose of common welfare. Such community is "Trinitarian", that is, it is an embodiment of reflexive and reflective purpose. In its reflexive consciousness it functions in-tensionally (of nature), and in its reflective consciousness it functions intentionally (of nurture). By reflexive and reflective agencies, consciousness motivates new outcomes that are inherent (in-tensional) and coherent (intentional). The power of edification, in-tensional and intentional, is quantum-electric, namely, *of body* (quantum) and *of soul* (electric), that is, it is at the same time material (embodied) and spiritual (energetic). Intention (spirituality) directs the construct (materiality) to become what it *intends*, namely, in accordance with the in-tensional laws of physical, biochemical interaction.

The noted astronomer Kepler (circa 1600) included a theological component to his exposition of Copernican heliocentrism, including the analogizing of the sphere to the Trinity. (Kenneth J. Howell, Ibid. pg. 128). To him, the sphere's center represented the Sun, the Father; the containing surface of the sphere represented the Son; and the communication within and between them represented the Holy Spirit. By analogy, this reflection may be applied to the atom, even though this analogy, like all analogies, limps—*analogia claudicat.* The nucleus at the center of the atom is the positive, gravity-point of the sphere, its ground state, the female principle, the First Person; the atom's spherical surface is energetically defined by the electrons, the male principle, the Second Person; and the electrical communication between them is the spiritual, holy energy that effects all manner of spiritual embodiment.

The experience of trinity in nature is as the agency of transformation, whereby and wherein two agents (spiritual/material) interact to bring about a new spiritual/material outcome. There is no change in consciousness (spirituality) except there is a change in materiality; the converse is also true. On reflection, one can understand how primitive observers could naturally come to the God-sense of Trinity because of the constant involvement of heavenly bodies in human experience. Genesis 1:16, "and God made two great luminaries, the big luminary to govern the day and the small luminary to govern the night." Manicheism reads in Genesis evidence of superior and "inferior" deities. [Kenneth J. Howell, Ibid. pg. 34]. The rising Sun measures the days in darkness and light—times for rest and activity. The Moon, the illuminated face of night, counts in thirty-day cycles its waxing and waning. Body cycles of sleep and waking follow the Sun's example, and woman's sexual waxing and waning (menses) follow the Moon's filling and emptying. From his/her perspective from Earth, the observer observes himself/herself to be the focus of heavenly display and enmeshed in the vital evidence of the Sun's, the Moon's and the Earth's doing. In some mysterious way it seems, focused as Earth is in the attention of Father Sun and Mother Moon, that they must be parents of Earth and life that springs from her. This tri-person community, Sun, Moon and Earth, is the constant agency of the times and events of human experience. How dramatically the evidence is written in the sky for all to read! How can human experience not come to a God-sense of Trinity when one is so totally enmeshed in the evidence?

It is the continuity experience of all vitality, especially in the case of the human family, that the origin of individuality and family is Trinitarian. Children are always person-made-flesh, a continuity processing evolved from dual parentage. Female and male, human beings as family, as community, are born of one spirit, of common "cosmic intention", even as soul/body is a unitary agent that serves purposes of personal and communal well-being. In the processing of self-aware consciousness, human insight reflects back on itself as identified in Godhead Personage; so, it is the evidence of conscious experience that the concept of Trinity Godhead is itself an evolved (inspired) consciousness of self-reflective insight that recognizes the inscription of Godhead in the work of the transformational cosmos.

The understanding of Trinity, as revealed by consciousness and faith, is a revelation of rationality. It is also true that the understandings of consciousness are themselves revelations of Trinity Process. Trinity Procession and process consciousness are essentially related. We are born into this Circle of Consciousness. While the mystery of revealed consciousness (faith and reason) is profound beyond comprehension, it is nevertheless true that *revealed consciousness* is a "natural" process. By natural processes of reflective consideration we can come to the understanding of being embedded in and developed from Trinitarian Spirituality.

The demand of public accountability alone can prevent professionals and institutions from appropriating reason and faith in self-interest; reason and faith are personally owned entitlements belonging to each individually. For the privileged role of contributing to life's *rising consciousness* we can only conclude that individual death is gain, not loss, for it is an essential aspect of spiritual/material transformation. Conscience compels the understanding that while human involvement in the natural order is an agency of change, human intention is not absolutely controlling, but rather it is accountable to natural laws, which ultimately prevail. In the hierarchy of God-ordained order, true worship necessarily includes self-conformity to the higher laws of transformational necessity, whose intension overrides human intention when push comes to shove. Network life embodies energetic refinements of original cosmic power and is the necessary subtlety in which humankind is embedded. Fidelity to truth is a consciousness above all others in the presupposition of worship, and the truth is that self-authentication and preservation wholly depend on the sustaining necessity of network life, which is therefore the minimal elemental object of awe pertaining to true worship. At the least level of energy/matter, human consciousness needs to acknowledge the hierarchical primacy of quantum-electric relationships in atoms and molecules, namely, the energetic (electrical) relationship between electrons and atomic nuclei that controls in all transformational processes.

God's revelation to us occurs in the unfolding process of creative complexity. God's Work, Creation, is a transformational cosmos, a "primary scripture" wherein divinity is scripted. God's image is inscripted in the open codes that manifest divine openness in the cosmos. Being born into this Great Cosmic Work, we are called to know it, love it and serve it. In reading Creation, we read God; in reading self-aware human nature, we read Self-aware Divine Nature.

Protestantism
Counter-Cultural & Reformational.

Both, in their meanings and in their historical expressions, "counter-cultural" and "reformational" can be read as synonyms; they are what the Protestant uprisings against Roman Catholicism were about. They are prophetic reactions against institutional abuses. Culture tends to become the complacent pursuit of the habitual; once habituated, patterns of complacency—routine—become embedded and difficult to change. They become the bedrock of institutional structures, and as long as they serve the institutional purposes, institutions will fight change. Institutionally and personally, we tend to like things to be predictable, manageable, conformed to our accustomed comfort zones. Today's crisis of the wholesale devastation of global biodiversity is troubling because it comes from settled but unsustainable patterns of institutional and personal consumerism. The institutionally caused havoc wasting global biodiversity—from the exploitation of species to the extinction and the massive waste of their habitats—is a creeping disease, long in coming, which roots in the human passion to subjugate, control and exploit nature for purposes of corporate advantage. The Garden-of-Eden conscience of male (human) dominion is not conscionably workable today.

It belongs to Protestantism to be counter-cultural, to challenge the *status quo* of cultural sin, but especially now, to challenge the sin of unsustainable, institutional havoc imposed on the networks of dependently diverse Earth life. The idolatry of institutional (corporate) primacy and complacency, represented globally in exploitive monoculturing, is a "religiously" passionate habituation intending to subjugate nature—presumed sinful and in need of redemption—to the hyped appetites of human consumption. Thoughtful people are more and more realizing that multi-national corporatism is a modern version of the "Tower of Babel", whose monoclonal voice broadcasts brazenly from its high ground and imposes its profiteering mind on global Earth's resources, even to the ultimate waste of natural diversity.

Institutions are not automatically credible; credibility must be earned. Institutions earn credibility by being consistently truthful and authentic. Truth and authenticity are internal/external expressions of inherently rational consistency—the external expression of work that conforms to an inner consciousness of the essential, global communication of Earthlife's interdependent networks.

Surely, as never before, these times challenge prophetic protest not only on behalf of human well-being, but also, on behalf of all Earthlife, for network life's vitality is the essentially dependent vitality of Eden's "middle tree". Network vitality is the "forbidden fruit" now being wasted so massively and flagrantly. Institutionalized theology registers an idolatrous complacency and has let it self be rationalized to serve corporate self-serving and mercantile consumerism. The exercise of prophetic religion demands that action be taken when conscience clarifies the need for action. Isn't the sin of institutional (Church) participation in global desecration sufficiently clarified now as to point the way for action against the unconscionable devastation of natural networks and resources? Individual conscience alone, in concert with other people of conscience, can effectively hold corporate mercantilism accountable for its sacrilegious waste of life.

There is in our time access to new insights in the codified word of natural scripture, not in man-written texts, but in the God-written text of *primary scripture*, the "quantum-electric cosmos". Interpreting the "light-code" of cosmic nature can open new insights into new theological understandings and "religious" relationships, that is, the relatedness of humans to each other, to nature and to all other life— which defines also the personal relationship with God. The science of Albert Einstein has uncovered the "rosetta stone" that contains the code of cosmic theology. All processes of cosmic transformation, of Earthlife vitality, are powered by a common dynamic, which can be defined as the "trimorphic resonance" of communication, consciousness and conscience. Faith roots in the ground of communication. Hope roots in the ground of consciousness. Love roots in the ground of conscience. The physical character of this *trimorphic process* and its outcomes of "virtue", in structure and function, are transformational phenomena originating in the "physics" of *harmonic resonance—molecular wave-energy.*

The reconciliation of faith-based communities with the verities of the *quantum-electric universe* may be the bedrock grounding by which a newly prophetic Protestantism can initiate a new reformation away from cultural embedded-ness in sacrilegious consumerism and toward the revitalization of biological diversification, the natural sacrament of network life's vitality.

Rising Consciousness

After the 09/11/01 terrorist attack and destruction of the World Trade Towers in New York City, the global community of nations is compelled to reassess prevailing business assumptions. While this assessment has already begun and already challenges a number of habituated but damaging presumptions, it is certain that even more changes in public consciousness will come about. "Quantum Religion" calls for a conscionable assessment of religion's acculturated presumptions, namely, that at the ground level, and below all else at the level of energy/particle relationships, that is, at the "quantum" level of the subatomic, atomic and molecular relationship and at the level of the logic/structure of quantum rationality, an inherently built-in system of consciousness exists upon which all societal structures can and must be built if they would prove to be sustainable. We must learn to work with nature and stop trying to force nature to conform to human brainstorms. We must calm our ego-centric, small-world passions and discover the "good news" lessons of "rising consciousness" that can be found in the living, indestructible spirit exemplified by the people of New York City following their/our tragedy. A global resurrection of consciousness, from an exploitive, one-voice determination of corporate consumerism to multi-voiced altruism and to the conservation and equitable use of resources, can come from the ashes of the World

Trade Towers if nations but open themselves to that saving possibility.

Rendering to Caesar

Doing like Rome, that is, building empires on the sweat of *slave* labor, is precisely what Christians and Christian Churches should not do if they/we would be identified by our "love for one another". "Romanism" as a societal model has been discredited by history and is not acceptable to modern consciousness. The whole sorry, colonial enterprise roots politically and religiously in European *Roman* empire-building.

That being said, Jesus also instructed, "Render to Caesar what belongs to Caesar, and to God what belongs to God." Clearly, Jesus seems to make the case that there is a secular, political order governing in matters of the public interest, and that we, as Christians, are obliged to find reconciliation with that order, even if, within it, there may be impositions (e.g., taxes) put upon us. Public well-being requires a certain amount of tolerance even under circumstances of personal disadvantage. It is in the nature of the love-relationship that the needs of others require us—Church—to participate in providing for the needs of others. Church is unfaithful to its public obligation if it becomes a joined agent in sustaining the unjust exploitation of people and resources. *Quantum religion* is about focusing consciousness on sustaining relationships that begin at the very deepest level of interaction.

Institutions of Church and State commonly labor under the same two handicaps, *ignorance* (making mistakes of misinformed judgment) and *arrogance* (refusing to admit errors of judgment). The Catholic Church as an institution is very firm in denying that it can err in moral judgments, even though Vatican II seems to signal a wholly different perception, specifically in the matter of collegial authenticity (including lay input) in the decision-making process of the Church.

The purpose of government is precisely to provide for the public good. It has no place in the business of allocating privilege, advantage and patronage, and inequitable access to public resources for personal and corporate advantage. The same should be said also in regards to other public institutions, such as, institutions involved in service work, of money exchange (banks), education (public schools and land grant universities), and, institutions that arbitrate actions between and amongst people and institutions (courts).

Violations of human rights and other forms of social repression are most violent when Church and State collaboratively sanction them. Such civil conspiracy occurs in theocratic governments, as, for example in the "Holy Roman Empire", and in our modern time in Afghanistan under the Islamic Taliban.

What has evolved in the United States of America (whose institutions in their roots are fed in Romanism) is a feudal system in which the institutions of government, corporate business, public universities, courts, and yes (!) churches collaborate in a way that preferential advantage, ownership, usage and distribution of land, food, mineral resources, etc., redounds disproportionately to an oligarchy of moneyed interest. The effect of this public corporate "conspiracy" is to sustain the profit-motive to the preferential advantage of public institutional insiders, that is, corporations, politicians, and lawyers, who largely control the adversarially structured political system. A large segment of the public has concluded that the Republican and Democrat Parties both serve preferentially corporate interests rather than the interests of the people.

Consequently, public cynicism toward the *pretenses* of legislators, namely, of pretending to be solicitous in the public interest while in fact being more focused on "where the money is"—for money is the *mother's milk* of politics, prevails. While the Democrat Party has seemed to be the Peoples' Party in past times, it is less distinguishable from the Republican Party in its corporate sycophancy. Used to further this "unholy alliance" are institutions of public education, which dutifully educate the *clerical* class of indoctrinated overseers of government, business, courts, and churches.

This describes the institutional fabric of a morphed feudalism that now prevails in Western culture, and which contributes to global turmoil and to the harsh divisions of world populations into the *few* "haves" and the *many* "have-nots". Global terrorism (*religiously* driven) is enabled by rampant injustices of resource usage and distribution.

Rather than allowing the disproportionate corporate exploitation of global resources, government should function as a "governor" of resource distribution and usage. Such *governance* rightly requires that bioregional resources not be exploited for the moneyed advantage of the few and to the disadvantage of the many, but, that the bioregion and the indigenes (not despots and despotism) benefit from the wealth of bioregional resources, and after that, to allow flow of benefits into channels that serve bioregional economics.

Presently, exploitation by corporate world trade seems to be ahead of the necessary political considerations of people-concerned governance. Old World, national colonial exploitation of populous "second world" countries has morphed in modern times into multi-national corporate exploitation. The terrorist bombing of the World Trade Towers in New York City is an unmistakable and shocking statement about the severity of this global problem that has through history wreaked devastation and stirred bloody contest.

What paradigm other than imperial/colonial *Romanism* suits the consciousness of contemporary global necessity? Sustainable society builds on nature's quantum-electric paradigm, namely, on life's evolutionary model that opts for relationships that are sustainable bottom-up from the minutest detail of energy/matter, soul/substance interaction. The continuity of the global *rising of consciousness* is sustained in *cosmic rationality* wherein spiritual resurrection, conscionable uplift, inherent in the dynamic of conscious transformation, *in fidelity to* the "natural paradigm".

Universal collaboration by global societies in sustaining bioregional diversity, equitable distribution and bioregional usage of resources, and the preservation of webs of floral, faunal, land and water interdependencies is the urgency of the present time. Life on Earth knows no borders except natural balance. With respect to the negative ecological impact of human presence on Earth, the sustainability of human life depends on humans exercising self-restraint in matters of the fact necessities of *natural balance* imposed by the evolutionary strategies of cosmic transformation.

Deciphering Godspeak

The alphabet of consciousness is composed of quantum-electric particles. All complexity of consciousness (spirituality) and structure (materiality) advances on the particulate suppositions of quantum-relativity. Every structure is particulately composed of clustered atoms of the same or different kinds. These are ciphers making up the "rosetta" codes of all spirituality/materiality, which reveal insights into relationships and introspective awareness. The imprint of divinity is revealed in this primacy scripture of cosmic consciousness, which is the paradigm of societal sustainability.

It is possible that professional science and professional religion are closer to reconciliation than either realizes. Perhaps reconciliation must take place first in public consciousness before these two professions break free from their ancient fixations; and perhaps it will first take place there. This is suggested for the reason that individual consciousness, the venue of public consciousness, is more and more discovering the universal language of universal relativity. And this is good, very good, for, for the first time public consciousness can expose the arrogance and error of the centrally entrenched fixations of science and religion, which speak a schizophrenic language and inflict on humankind a disease that is fatally uprooting unless remedied early enough.

"Vox populi, vox Dei." The people's voice will not be silenced, nor will it be disregarded. If science and religion ignore it, they will be exposed for their ignorance, arrogance and bad faith.

The Unnoticed Elephant

IN PREVISIONING A HARMONY OF THE GLOBAL FAMILY OF NATIONS, WHICH NOT MERELY TOLERATES OTHERS, BUT WHICH LOVES ALL, THE QUESTION IS FITTINGLY

ASKED, "IN WHAT WAYS MIGHT CHRISTIAN RELIGION CHANGE, STRUCTURALLY AND THEOLOGICALLY, IF IT WOULD COME TO UNDERSTANDINGS THAT ARE PREMISED IN THE QUANTUM RELATIVE WORLDVIEW?"

Without pointing a finger of blame at any Christian denomination, it seems fair to observe that "quantum relativity" is the *unnoticed elephant* in virtually every Christian sanctuary. Before suggesting areas of institutional and theological change in Christian denominations, a clear sense of the meaning of "quantum relativity" needs to be put in place. *Quantum relativity* is the energetic relationship responsible for the structuring of all matter, including Earthlife itself. The substantive structure of all matter is atomic; the particulate matrix of atoms is a complex of charged particles. The inherent cohesion of atoms is "quantum-electric"; the subatomic substances (quanta) composing nuclei and electrons, the elemental structures of atoms, are *quantum-electrically related*, that is, they are inter-relationally and inherently dynamic by reason of particulate attraction/repulsion, according to their electrical charges. The disposition of energy in matter (particulate and atomic), the structure of matter itself, is quantum-electric, is quantum-electrically *relative* (related). Matter is the effective and efficient quantification and qualification of energy; matter is energy packaged in usefully creative "quanta". Einstein says that matter is convertible back to energy. Herein is stated the fundamental, universal connectivity, relativity, of all substance, all structure, based on the common energetic intension (spirituality) of all matter/structure.

The essentially transformational nature of all energy/matter originates in, and carries forward from the intension of universal energy/matter, the adhesion of positive/negative particles. The egalitarian character of all energy/matter-substantiation, its quantum relativity, is for all time cosmically disposed and qualified. The complex interactions of substantiated energy, characterized by quantum relativity, account for the subtle and complex expressions of energy (*soul*). In a broad and inclusive sense, primal energy, relationally qualified in all matter, may be understood in its fundamental nature as the "spirituality" of matter. All matter, all structure, possesses "spirituality". In the cosmic sense, spirituality is itself *quantum relative.*

The *matter* of Church (religion), in the quantum-electric analogy, is its "structure", and its *energy* (spirituality, soul) is its theology. Institution/theology (substantiated energy) is the perceived matter/energy of Church, of established religion. If this is true, then, the common basis for all authentic religion is cosmic in origin and in its historical expression. The authentic structuring/expression of Church depends upon its fidelity to the authentic expression of quantum relativity in cosmic creation. Misinformed consciousness leads to misinformed substantiation (structures) and flawed relationships that frustrate social harmony and sustainability.

Continuous transformations of universal consciousness, based on experience in the changing cosmos, account for changes of human insights, thus, in the course of time, it seems inevitable that changes also must occur in Church structure and theology. While I venture to suggest some needed changes below, I do not suggest that they are necessarily the most important ones, much less, that they are the only ones.

1.) *A theology of the equality of the sexes.* In the elemental structuring of the atom, the electron cannot say to the nucleus, "I am more important in our relationship than you. I have been elected by God to perform a superior role in our relationship." Neither can males truthfully assume such claim over females.

2.) *The hierarchical relationship.* The complexity of intensional/intentional spirituality predisposes substances and motivates them toward diverse transformations and sequentially dependent relationships. The "intentionality" of relationships is expressed in symbiotic accommodations that beneficially serve sustainable interactions. Authority, or effective power in human relationships, is authentically expressed when it is based on a symbiotic consciousness that is more perfected in its insight (prevision) into beneficial interrelationship, whether the consciousness comes from a male or female person. Higher relational consciousness (knowledge, experience, judgment) is a more effective expression of authenticity in socially ordered relationships than is sexuality. It is patently wrong to structure societal hierarchies on the illogical and wrong culture of belief that female intelligence is inferior, and that females are less "chosen" by God than males. Before God both sexes, all races and cultures have equal standing.

3.) *The "sacrament" nature of all relationship.* Structure (matter) is an information-loaded "sign" of the inner energy (intentionality, spirituality), which disposes matter/energy in *particulate* definition. All of creation is a unitary, connected, "mystical body", expressed in common resources. Even as the human body (male and female) is relationally differentiated in its various organs and their characteristic functions, so are all the energetically disposed bodies in the cosmos.

4.) *Consciousness of God.* Even as cosmic energy/matter continues its impulsive self-transformation into greater subtlety and complexity, so too does consciously human spirituality. Necessarily, the human conception of God is a changing conception even as human consciousness changes with complexity's continuing transformation. New understandings of divinity, Godhead, arise in the self-consciousness of self-unfolding in cosmic relationships.

The arrogation of dominion and the culture of destructive alienations by peoples against one another cannot be defended on presumptions of religious authenticity. Quite to the contrary, the violence of domination and cultural alienation, of religious elitism, roots in misperceived and misguided understandings of naturally essential relationships. The only code for constructing a sustainable world order is written in natural scripture, the primacy word of the unspeakable God. As a global people, we must decode this natural scripture and live religiously in harmony with it, if we are to be faithful to God's word and advance a sustainable, human presence on Earth.

Sylvester L. Steffen

The Unforgivable Sin

Hard headedness, fixation, incorrigibility, the refusal to be open to new learning, are what may be characterized as "sins against the Holy Spirit". Cemented in the intransigence of arrogant opinion, we commit ourselves to the fate of becoming burned-out *lucifers*. Because incorrigibility is an arrogant hardness of heart and mind it fits in the category of sin which Scripture calls "unforgivable", the sin against the Holy Spirit. (Mt 12, 31; Mk 3, 29; Lk 11, 15; 12, 10; Hb 6, 4-6; 10, 26; 1 Jn 5, 16.)

The Holy Spirit is the Divine Breath that inspires and empowers every least cosmic particle, and which vitalizes all Earth consciousness. We are born into the grand panoply of diversified life and we are sustained by it; spiritual vitality is invigorated by the infinite display of divinity. In the seamless continuity of life and consciousness, all nature flourishes by the powers of the same Holy Spirit, whose voice speaks a unique language within each living creature. When a species of life is destroyed it represents a final silencing of a specially refined spirituality and a lessening of inspiration in Earthlife. Because species-extinction is irreversible, it means that there never can be a return of the lost species. The finality of the loss is unforgiving; it is an irredeemable act of loss, an in-your-face rebuke to providential presence.

When we trash Earthlife, we trash some uniquely sustaining presence of divinity embedded in the Sacrament of God Present in natural providence. The global sufferings of humankind are not from a lack of nurture in nature but from uncurbed appetites that wastefully exploit providence and the Inspirational Soul of life. Each wasted species represents a sin directed against God. When God's voice in a species is extinguished so is the special grace that has been perfected and projected by it throughout the links of network life. Whole chapters of God's Earth Work are being wiped out on a daily basis by human greed and mindlessness. If this continues unchecked what will survive after such cannibalism? Havoc and desolation, unless life's surviving remnants can be salvaged and resurrected.

Today's excessive violence visited on nature by humankind is more far reaching than any other visited before on Earth by Earth-creatures. The fearsome malice of this destruction is that it is an act of human mindlessness, not an act of God. It arises out of the obsessive passion to control and possess. Every species extinguished, whether by consumption, by bombs or by chemical toxicity is an effect of human willfulness. In effect such wrongdoing is an act that frustrates the Creator of the Garden of His/Her making. This wasteful doing, the work of ego-motivated exploitation, is in all its aspects a kind of idolatry. The global magnitude of the ongoing sin of idolatry should strike fear in us. And so should the realization that you and I participate in the sin if for no other reason than that we are born into the social habituation of it.

However, we cannot be faulted for occurrences that precede us. So, we should take heart and realize that by our seeking to change we can facilitate redemption and movement away from past bad habits. Because of the havoc wreaked on it, the Earth has suffered harm and continues to suffer harm, but it is in our power to engage the natural resilience of Earthlife and facilitate its redemption. But before we can become agents of redemption we must first recognize in the specific the hurt of nature and acknowledge the causes of it; only then can needed restitution be assessed and repair begun on the frayed fabric of network life. Virtually all life on Earth is at risk, so, no one, no matter where in the world one is, needs to look any further than the circumstances at hand to discover needed remediation and restoration. Failure to engage oneself in this tasks of remediation and reparation is in the least failure of insight if not willful implication in perpetuating the havoc.

The agency of redemption is first and foremost a mindset. Firm purpose to amend one's way of living isn't likely to happen fast and easy, nor will the repair of Earth havoc. Purpose of amendment may (should) arise out of one's personal recognition of the moral obligation to change one's personally injurious habits; it isn't likely to come about without such recognition. In this confessional work of Church we can and should all participate. Such confessional work is an essential aspect of authentic worship, communally obliging.

Cosmic Paul

From the perspective of Quantum Theology the writings of Paul take on new dimensions of understanding. For example, in Chapter 8 of First Romans, Paul deals with the body and spirit of human life (verse. 1). Christ consciousness liberates human consciousness. Christ consciousness frees us from the apparent futility of death and human failing. The narrow insights of individual egoism imprison consciousness. (verse 2); the narrow defines of individuality make one powerless except one taps into the field of cosmic consciousness—*Christic* (verse. 3). (The phrase "life in Christ" may be read in the sense of cosmic consciousness.)

In the Christ-experience, consciousness recognizes divinity in "Son-ship", generation after generation, becoming ever more aware in the consciousness of each successive generation. The "just demands" of self-aware consciousness are fulfilled not by obsession in body pleasure but by the ever-expanding consciousness of spirit (verse. 4). If we would expand our consciousness and grow in the fullness of life we must connect our consciousness to intuitional wisdom, the open rationality of cosmic consciousness. The giftedness of the body is its spiritual power, operative in every least cell and in the unity-purpose of body members (verse. 5). In transformational consciousness we know that materiality changes according to cosmic laws. Our bodies are subject to these laws which require *the seed to die* in order for new life, new consciousness to arise. In self-aware consciousness, cosmically rooted, we discover personal fulfillment, peace (verse. 6). If we obsess in preoccupations with materiality, which is transitional, we frustrate the deep consciousness of Word advancing in Spirit awareness. Godlikeness is in spirituality; preoccupation in fleeting pleasure of the body frustrates spirituality, Godlikeness (verse. 7).

Christ consciousness, reflective consciousness that preoccupies itself in the ocean potential of spirituality, is where we find peace and fullness. The true Christian—the cosmic aware person—is spiritually aware beyond flesh concerns (verse. 9). The law of material *corruptibility*, of transformational necessity, frustrates consciousness if one is fixated in materiality. Preoccupation in materiality is a bad choice (sin), whereas, the body is *justified*, validated, in the spiritual power and potential it possesses by virtue of its cosmic giftedness (verse. 10). As we discover in the resurrection of Jesus the divine uplift of consciousness, so we will contribute to the divine uplift of consciousness by our living "in the Spirit". Our indebtedness is to the torch of spirituality glowing within that empowers our consciousness and forges body transformation. If we choose to live committed to spiritual enlightenment, we will live on in the deeds that we do, in the love we have for others, for example takes root in their lives (verses. 11-13).

Our "birth in spirit", our cosmic inheritance, identifies us as divinely connected (verse. 14). Birth into the spiritual openness of cosmic consciousness liberates us from the enslavement of old fears and empowers us in the knowledge that we have a well-tested Parentage (verse. 15). Cosmic spirituality, whence we arise, testifies to our Christic Personhood (verse. 16). If, in Christic Personhood, we enter into the passions of transformation and embrace the ultimate giving up of our bodies, we will share cosmic glory with Christ (verse. 17). From our present perspective we can glory in the cosmic consciousness we have inherited, incomplete as it is, and know that it is within our power to add to the expanding glory of cosmic consciousness, into which future generations will be born (verse. 18). Indeed, the cosmos anticipates the new revelations of consciousness that will come to fruition in generations ("sons") to come (verse. 19).

If our expectation of cosmic glory, yet to arise in pooled consciousness, reaches the level of conviction, then, indeed, the sufferings we experience are nothing by comparison. Creation must obey the laws that empower it; embedded in the fabric of creation, we too are subject to its laws. Consciousness that fails to arise above the daily humdrum, above ephemeral obsessions with the body, is futile, indeed; but, we possess hope, the awareness that we are empowered to bring cosmic consciousness to glory (verse. 20). We know that future generations will be even less enslaved to the futility of corruptions by virtue of the options of spiritual choices we make (verse. 22).

It is in the nature of Earthlife always to be in the process of birthing (verse. 23). Not just (mother) Earth but we also labor to keep consciousness focused on the redemptive power, which is the "first fruit" of spiritual commitment. In the exercise of hope we enable redemption, the option of well-being that serves not selfishness but selflessness, the witness of love in which we were born. Though we ourselves may not witness outcomes of good that result from our exercise of hope, we know that such outcomes will come to be. We have endurance in the knowledge even of what we cannot see (verses. 24, 25).

Even as the origin of our consciousness springs from the well of cosmic consciousness, so also that same well sustains and refreshes us in our spiritual resolve. This well is material and spiritual, genetic and memetic; spiritual empowerment is an *aura*, an energetic field that envelops molecular materiality, small and large, in which energy-spirit resides. This spiritual power mediates and sustains the fruits of the travail of *groaning* materiality (verse. 28).

All are called to live within the purposeful, sustaining power, which makes all things "work to good" even when transformation requires material dissolution. Many of life's changes are cyclical, repetitive and predictable. These present little challenge to consciousness. But, unpredictable events, some of which will be cataclysmic, enter our experience and challenge us to the core. Our conscious reserves and spiritual resolves are stretched to their limits. Sickness, deaths, tragedies, but also experiences of success and accomplishment can profoundly change us. All can work to good. Since the destruction of the twin towers of the World Trade Center in New York City, a global change of consciousness is occurring. It is likely that global consciousness will not return to the *status quo* that prevailed before that event. Even this event can work to good. The cosmic spirit groans to empower materiality to achieve the glory which spirituality envisions and foreknows (verse. 29).

All are called to the same *justification* that comes with the willing and witting spirituality of cosmic consciousness. A life lived in spiritual justification is a life destined to share in the glory of justification (verse. 30). Justification is growth into right-mindedness that comes with the resolve to live rightly, harmoniously, with sensitivity for common well-being. In lived justification we choose God, the sacred mind in all that matters (verse. 31).

If like Jesus we endure in the same spirit of harmony with God's Word, we are assured of being granted the resolve that we need to deal with whatever comes our way. By our endurance, our own justification becomes the justification of Christ Jesus (verses. 32-34). No power can separate us from the love of God that comes with Christ-living and Christ-consciousness (verses. 35-39).

As consciousness resides in quantum bits, its ascendancy is empowered and qualified in them and in their subtle connections. In the quantum differentiation of bits, consciousness rises. As matter becomes more complex in the joining of quantum bits, so does consciousness. Complex consciousness is energy subtlety tensioned in molecular arrangements. Soul like energy is prior spirituality; soul is qualified energy of prior spirituality and is defined in its structural particularity. Prior Spirituality is the infinite sea of conscious potential, the Prior Spirituality implicated by divine intention in the pre-Big-bang point-density. The return of soul is return to God-implicate Spirituality. Divine Implication qualifies soul's origination. Consciousness is awareness-growth in matter, the reciprocal purpose of joined energy/matter. "Justification" is a matter of reciprocal purpose wherein joined energy/matter is proven by/in sustainable intension/intention. Sustainable diversification is a law of cosmic success, of inter-relational success. It is a condition of intended purpose. Failure of intent is failure of purpose; failure of purpose is failure of sustainability. Degradation and collapse are fueled by wrong intention and by the impetus of consumptive excess; wrong intention and excess trash the fruitful sustainability of the "middletree". Wanton consumption and excess obsession in self are sins against *middletree vitality*. Cannibalistic self-consumption is original sin.

In our time, extremist Islamic fundamentalism, in the violent pattern of Cain in hatred and rage against his brother Abel, now plots murder and destruction and violates God's expectation of justification. The violence of intolerance, born from excess self-obsession, devastates justification. Self-obsession polarizes the extremes, and the violence of extremes instigates the destruction of the "Garden" and wastes its essential vitality. Capitalist consumerism, associated with Christianity, is a pole apart from the Islamic life-style. The polarities of life-styles are associated with religious animus. The violent wars of vengeful, "religious" intolerance trash the Garden with uncommon malice. When will Adam's children ever learn? The answer, and the future hope is "Christ consciousness"—*Christogenesis*—living not only non-violently but also with unconditional love for one another, including love for one's enemies. Justification is not an Adam-consciousness but it is a Christ-consciousness, ascendancy to a "new phylum of love".

Flat Earth Theology

Religious consciousness struggles with perspective, with giving proportionate valuation to relationships that are "transcendent" and "immanent"—relationships that are "from above" (between man and God) and "from below" (between man and man). For example, in Jesus' person, how do we reconcile/understand divine/human relationships as they obtained during his historical presence on Earth, and how they obtain in the world today, absent his physical presence on Earth. The difficulty of proportional valuation comes with discernment of how divinity (*transcendence*, that which is *from above*) and humanity (*immanence,* that which is *from below*) function fully and harmoniously in the person of Jesus, then and now.

Within the Church there has been an inside tug-of-war between two theological poles trying to give balance to the interaction between Jesus' divine nature and his human nature. And it continues. (See COMMONWEAL, "Christology, What it is, and Why it is Important", contributors, Robert Krieg, Terrence Tilley and Sarah Coakley, March 22, 2002, Volume CXXIX, Number 6, pp 12-17, Commonweal Foundation, 475 Riverside Drive, Room 405, New York, N.Y. 10115).

Linear theology evolved essentially from a presumption of Earth/human centrism and the God-man relationship in the centrist cosmology. In flat-Earth consciousness descriptions of relationships tend to be from point to point—straight line—for example, from God to human, human to God; and on Earth's flat plane relationships are horizontal. So, linear relationships are "vertical" and "horizontal". On the horizontal plane, moral, religious relationships are those between and amongst humans, *man-to-man*. Notably missing from the equation are moral relationships of humans to "other", to those essential and common dependencies upon which all rely, namely, food, shelter, drink, etc., and whence they come, namely, from Earth's ecology.

The problem arises in considering the proportional rights of individuals to the commonly necessary dependencies. Where is the *place* of morality, religious connection, in the consideration of these relationships? The dimension of the moral relationship in the use of subsistent necessities is missing in the simplistically linear casting of theology. Natural complexity and its problems measure beyond cosmological/theological linearity, and is a phenomenon of "quantum cosmology", whose essential dimensions include time and space in addition to energy and matter. In the natural order, "time" and "place" (space) fit in the ethics of relationships and have moral weight, that is, they have consequence in and on human relationships.

Place—network, Earth ecology—is the common venue of conscious human and divine presence. This venue is cosmic, quantum-electric, Earth situated. Humankind (Church and State) is morally obliged not to waste network life and balance and not to suffocate by prolix consumption the vital sustainability of the human/divine venue. Hyper-ego assertion is an idolatrous arrogation against divine order and priority.

Linear theological consciousness fails to encompass the expansive and multiple dimensions of moral obligation. The *time* of moral obligation is "imminent", immediate, now and always, this moment and every moment. Consequences of relationships, considerate and inconsiderate, accumulate to the good or detriment of common vitality, ecology, the venue of common *place-space*. Moral obligation pertains to the common, conscious obligation of all to anticipate and avoid behavior that is destructive of the locus of all living relationships.

The problem arises in the proportional balancing of the consciousness of interdependency and the usage, personal and social, of ecological necessities, material and spiritual. The theological issue at the core of this conundrum is "Where do we meet God? Where do we meet Jesus? And, where is Jesus in the real world, today?" Immediate answers are: We find Jesus in one another. We must do to one another as Jesus did. We find God everywhere present. We find God in person-to-person relationships. And we find God in Nature's Spirituality. We find God in natural providence. God's Spirit is Soul of the Universe. God's Spirit is Cosmic Energy, the Soul of all Vitality—Divinity Present—moment-by-moment. By right of birth and baptism we participate in benefices and obligations of divine inheritance.

If this is true, then we must conclude that until now, Christian Theology has been inadequate and ineffective, and is now inadequate and ineffective in motivating people in their times to grasp the immediacy, the imminence, of divine presence, to them and to us in The Here and Now.

We fail to account for the real imminence of divine presence in space/time, and how this moral failure impacts the essential continuity of space/time-dependent reality. The Presence of Trinity is real, in cause and effect, in the transformational universe. Process Trinity, in structure (sign) and effect (grace), amplifies and energizes natural transformations that are ongoing by virtue of the resonance of harmonized faith, hope and love—amplifications of divine/human communication, consciousness and conscience.

It is only in the consciousness of the dimensions of the quantum universe and their moral interdependency, as qualified by human relationships that we can begin to understand the meaning of "quantum religion". Quantum religion finds its wings and breaks free from the misdirection of theological linearity when it discovers the potential expansiveness of relationships in cosmic space/time dimensions.

Quantum Religion/Relativity

St. Anselm's perennially quoted line "faith seeking intelligence" has the bride pursuing the groom. But, it needs to work the other way also, namely, that grooms pursue their brides. And so it is between science and religion. The intuitive wisdom of consciousness (Fides) pursues new knowledge (Ratio) even as new knowledge must accommodate to intuitional wisdom.

The fabric of life itself suggests a "string theory" for everything. For example, memory patterns become enduring threads in the genetic/memetic fabric of self-awareness. Each strand identifies with the conscious whole, which is self-renewed in the weave of body/soul experience. The unity of conscious fidelity is in strand integrity, which originates from and finds its identity in mind/body wholeness. Integrity's grounding is faith-consciousness, the secure awareness of strand identity fitted within the weave of the seamless whole. Reason seeks understanding of fidelity even as fidelity is secured in the understanding of reason. Their bonded authenticity vitalizes the integral whole, which unfolds from and is transformed by their communion. Religion and science grow together in a dance of mutual consciousness, which is made flesh in Faith/Reason's renewing fabric.

Quantum religion enlightens the natural turf of Faith/Reason's holistic rationality, wherein/on they mutually exercise their nuptial enticement and consummate the renewal of religion and science. The calculated divorce of institutions of religion and science has proven to be devastating for both Faith and Reason, and for their offspring, the sibling institutions of religion and science.

Quantum Religion is connection—binding at the least level of being and at the highest. Connection is a consciousness of purpose. Purpose is contagious. It is passed on and on. Purpose gives growth to consciousness. We are born into conscious connection—purpose. We are destined to growth, to rising, and to resurrection. Consciousness propels rising, growth and resurrection. With growth, least things become greater things. All purposely intends to greater growth. This is our destiny. It is expansive. It is wild. It is open. It is inclusive. It reaches out. It is bubbly. It is creative. It doesn't stop. It is destiny with greatness. It is something *familiar*. It's about our deep self. It's, "I've been there before!" It's *deja vous* all over again! It's about the certitude of *relativity*.

I don't know who coined the term "quantum relativity", but Albert Einstein has given it wings. Even though the term is widely used, it is probably not generally well understood; certainly not in terms of its deep connections to everything we are and do.

Put in terms of "quantum religion", we may perhaps better connect with meanings of quantum relativity. Quantum relativity is the deep theology of quantum religion. *Naturally*, it is what and who we are. It's what life is all about. It's what the sense of being free and connected at the same time is all about. It's what truth is about, your truth and my truth. Who lives by fear does not have his own truth; he lives by another's truth. Freedom is the power of deep understanding. Go, seek and find yourself. Knock. The door of self is open. It is good news. It is an invitation to Godlikeness. It's the discovery of Godlikeness. It's your choice. Hang on. Let it grow.

Liberation Theology

The accelerated pace of global impoverishment, governmental crises, wars and terror in the twentieth century leave little doubt that the old answers of economics, politics and theology are not adequate for dealing with the ever-increasing complexities of the global reality. Two World Wars, the spread of global Communism, national turmoil, violations of human rights, people unrest and the convening of the Second Vatican Council testify to the global disjointing of world populations.

The imminence of burgeoning perils has made clear the explosive character of the depletions of space and of time running out, economically and ecologically. New science confronts old faith; democracy confronts despotism, and Liberation Theology (open and inclusive) confronts the linear, centrist, closed and exclusive theologies of the past. Women are telling men loud and clear, "Boys, your old games are over. They don't work!"

The birth of Liberation Theology occurred specifically in those places, which suffered the greatest impoverishment from the smothering spread of *Catholic* Colonialism. Impoverishment has occurred in areas of basic food and sanitation needs from population pressures and from the collapse of indigenous cultures due to outside overreaching. Old religions and old governments are being swamped by a mass tide of global suffering and terror that have grown out of colonialism's mutual sins of unjust domination and exploitation. The sins of the past have come of age.

Pope John XXIII had both the intellectual insight and the moral courage to confront "the people" Church with the global mess and with the responsibility of rising to the task. From the inspiration of Vatican II, and from intellectual and moral awareness, Third World Theologians, in the midst of their downtrodden people, have arisen and now offer a new theology that goes beyond the linear dimensions of old theology and deals with the real time/space crises of people globally. The imminent realities of desperation demand immediate attention. Rising on the crest of this fast growing tide are innovative theologians, especially from Central and South America, who are giving attention to the inadequacy of simplistic theologies and politics. Their company includes, Benedictine Bishop Dom Helder Camara, Recife, Brazil; the Uruguayan Jesuit Juan Luis Segundo, the Peruvian priest Gustavo Gutierrez, the Brazilian Franciscan Leonardo Boffo, and the Basque Jesuit Jon Sobrino working in El Salvador. (See Dickinson, "The Dialectical Development…" Pg 300.)

What is fundamentally new in Liberation Theology is the interface of politics and theology, for, in today's world reality, the moral crisis is the exploitation (religiously and governmentally sanctioned) of indigenous peoples and resources by the few, and the deep impoverishment of the many. Economics and Ecologics have been put to cross-purposes by the philosophy/theology of the politics/religion of colonialism, so that now economies and ecologies alike are in shambles and indigenous people have been reduced to "Third World" desperation.

It's been said that, "Everything is political." (Ibid: pg 304). Another way of saying essentially the same thing is, "Everything is theological." In the quantum-relativity perspective, spirituality and materiality are a joined continuum in which politics and theology coalesce. Matter (politics) is energy (spirituality-theology). Body is soul-energy. The human reduces to the divine; politics reduce to theology, and economics to ecologics. Ecology and economy are only sustainable in the diverse webs of living systems harmoniously supporting each other. The Church and Secular Society are One People. The People are Church. The People are State.

Earth is the "place", and the "time" is now, where and when *quantum religion* and *quantum relativity* must make their peace, in terms of new theology, new politics and new economics.

Protestantism arose at a time of the dawning of new science, new technology and new moral consciousness, and confronted old science and old church. But the conspiracy of centrist Church and centrist State objected and obstructed their protest. Liberation Theology gives new Spirit to authentic Protestant and Catholic insight at a time perhaps more critical than the time of their original parting. And in their moral mission of peace and justice, both seem to be at the cusp of healing their old divisions and of bringing new hope to a sorely suffering world. There will be peace when economics and ecologics are harmoniously restored, and when Church and State are reconciled in the common purpose of cosmic consciousness, pursuing global (ecologic-economic) peace and justice.

Reason Searching Faith

"God made two great lights, the greater one to govern the day, and the lesser to govern the night, and he made the stars. God set them in the dome of the sky, to shed light upon the earth". Genesis I 16, 17

Reason doesn't exist alone. Like consciousness, Reason expresses, and is expressed in, mutuality. From early times, reflective self-consciousness identifies divinity as Light, and the luminaries of the sky, the "greater" and the "lesser", to be of divine origin and purpose. Light, whether emanating from the Sun or reflected in the Moon's face, is one and the same Consciousness—Faith/Reason searching Sense in Nature. Faith is sunlight (processed reason) beaming from the Moon's face.

Light's original searching begins with Word. Light and Word together is an Agency of joined vision, perception and insight; and in the process of mutuality, Reason (Light) and Faith (Word) are bonded in common, holy Purpose, and are inspiration in Earth substance, whose soul is Love's Inspiration—life's Alpha and Omega.

Quantum Leap in Consciousness

Consciousness is an arrow in time; communication produces it and propels it by advancing it on the prior and ongoing insights of experiential dialog. Suppressive conditions can repress consciousness, subvert it, and/or frustrate its advance. However, upon being liberated from subversion/frustration, consciousness may advance in quantum leaps. Experiences of such quantum leaps are evidenced in history.

By birth and by culture, human beings are coins of two faces, expressive and impressive. The genetic codes of birth destine (express) the unfolding of personality to follow natural codes of genetic imprint. These codes (DNA) are forged by fire and in the mold of evolutionary experience, the trial-and-error caldron that tests the rigidity and pliability of personal, conscious embodiment. Opposite the expressive face of human character is the *impressive*. It's as if the coin of personality is a stamped image whose likeness is preserved as two faces, but in reverse of each other. The genetic codes of nature give the human individual an assertive face that stands out on the one side but which is indented on the other, and which works as a cup collecting and processing messages of experience.

The times of history are sometimes more impressive and sometimes less, that is, sometimes heavier and sometimes lighter. The ordeals of historical experience are sometimes so overwhelming that radical shifts of culture occur because of the unusual gravity of experience; such shifts are precipitated by radical pressures, which force the "impressive" side of consciousness to flip from its passive face to its active, "expressive" face. When the public coin of consciousness flips from its *impressive* to its *expressive* side a sea change may occur. A sea change shift of paradigms may now be occurring.

In our lifetimes we are experiencing shifts. Prior to the Protestant Reformation, the cultured faith of Christian religion was shaped by a hyped consciousness of faith. The general public was cultured ("impressed") to accept belief in what church authority, and state, told them to accept. If individuals deviated from cultural belief, authoritarian control mechanisms were put in place, which made the deviants pay dearly—even with their lives. The "peace" of this culture was a physically exacted, repressive peace. The gross malfeasance of religious/state authoritarians showed its brazen face, and the public responded against the arrogant violence of the authoritarians.

The reactionary battle against the gross arrogance of fideistic oppressors in Europe brought about a public shift to *expressive public consciousness* that sought to affirm a restored balance between faith and reason, probity and license; this shift was The First Enlightenment, which followed the Thirty Years Wars of Religions in Europe, 1618-1648.

Fideistic religion, deeply rooted in Old Law tradition and in some 1500 years of New (Testament) Law, was unwilling to surrender its grip to the new wave of rationality that was insinuating itself (humanism) in virtually every social arena of European society. Holy Roman Catholic (imperial religion) mounted an uncompromising, bloody, counter-reformation offensive, in the forms of Inquisitions, Spanish and Roman, Witch-hunts and Church Councils, specifically, the Councils of Trent and the First Vatican. Torture and death by burning at the stake were devices used to preserve and promote orthodox belief.

The counter-reformation Church Councils also refined faith dogma, thereby being able to ascertain more specifically when citizens were heretical in their beliefs and in violation of church/state law. The constraints of faith on reason impacted with greater impression on the face of public consciousness. Catholics educated in the faith as it was defined under counter-reformation initiatives are referred to as "Tridentine" (Council of Trent) Catholics, for their faith wears the imprint of rigid dogmatic definition.

Tridentine Catholicism persisted rigorously right up to the mid-1960s, when Pope John XXIII convened the Second Vatican Council for the objective of "aggiornamento", *updating*. The very anticipation of the Second Vatican Council created an air of electric release from the charged fideism of the prior two Church Councils. Indeed, the Second Vatican Council ushered in an era of Catholic consciousness decidedly away from a repressed lay consciousness to an open expression of lay activism—away from the hyped domination of fideism to a balanced relationship between faith and reason.

My religious upbringing in Tridentine Faith took place from my birth in 1933 up to the 1960s. As an eleven-year student (1946-1957) to the Catholic priesthood, and a student of theology in 1957, I was required to write a paper on a theological subject of my choosing. Being caught up in the electricity of the time, the zenith time of Tridentine fideism and the restive atmosphere of rational stirrings, my paper revealed me to be a person between times. The topic I chose was "RELIGION: A Rational Consideration". (See: APPENDIX A, Pg. 255). In reviewing this writing, forty-five years later, the tenor of the paper is more understandable in the context of *rational stirrings* within Catholicism at that time in my life. What seems evident to me now is that I was struggling to find my own consciousness, my own rationality—not that I rejected the faith of my upbringing but that I was trying to find correspondence between the *impressive* consciousness of my upbringing and the *expressive* consciousness of my own intellectual enlightenment.

The radical shift that occurred (is occurring) in religious rationality, as the result of Vatican II, is only now taking serious issue with the hard rock consciousness of counter-reformation faith. Public consciousness is now making a sea change away from the harsh constraints of *impressive fideism* to the *expressive rationality* of enlightened consciousness. This shift is so profound and so broadly distributed as to leave little question that this generation is moving toward a new global cultural era of Second Enlightenment, in which reason is asserting its co-equal place alongside the hyper culture of dogmatic faith.

Zeal for Life.

Zeal for life is a divinely inspired grace. The channel of this grace's working is conscience-sensitive, and when offended, drives to defensive anger. Disrespect for life aroused Jesus' anger with the moneychangers in the temple. Money-changing religion desecrates life. Zeal for life goes to the essence of parental conscience, of man's respect for woman, of woman's respect for man, and for their mutual respect for the life they beget.

The angry and uncompromising response of lay women and men to the sexual abuse by clergy is righteous because the abuse disrespects divine sensitivity for life at its incipient and essential base, the delicate love relationship between man and woman and their mutual responsibility to nurture, sustain, honor, and love least life. The chagrin, hurt and anger over abortions of defenseless life are righteous, called-for and necessary. The seamless nature of love, life, and specifically, love for life, calls for righteous response. Response is honorable and righteous when it is authentic, not ideological (pretentious), when it is consistent, not selective. The selective focus on one aspect of right to life, in such a disproportionate way as to blind conscience toward other equally important aspects of life is inauthentic. Heresy is rightly described as "excessive zeal for one truth that violates other truths".

Respect, zeal for life, must be a consistent ethic that is equally sensitive to all right-to-life causes, for the right to life of the unborn child, and for the right to life of the divine order by which life is nurtured, sustained and honored, namely, vital sexuality and life's global ecozoic (life-dependent ecology) network—the continuity basis of all life, the <u>naturalis</u> <u>sacramentum</u> <u>ordinis</u> that nurtures, sustains and honors life globally.

It is inauthentic, for example, to be fanatical for one cause and blind to other right-to-life causes. Fidelity to Catholic theology's

Consistent Life Ethic requires of each of us to have zeal also for the causes of clerical responsibility (sexual) for women and children and ecological responsibility toward network life, for in the ethic of proportionality, these right-to-life causes have moral standing with anti-abortion causes. Each requires equal conscience; their causes must be joined.

ALPHA GENESIS

The Blessing Wash Of Crystal Water
Overflows The Earthen Vessel
And Enlivens Earth With Miracles.
Unbruised, Vintage Water Sparkles Lively Drinkers.
Water Works To Transform Creation
In Unending Celebration Of Wedding.

Woman, God's Mother,
The Inspired Well Of Baptismal Giving,
Hallows The Family Way And Continues
To Order Up Divine Participation
In Celebrating Newlife Wine
Within Everfresh Placental Skins.

SECTION II

ON FAITH

FAITH is about the certitude of "things unknown", of things that are beyond the competence of human knowing. Faith is a faculty of knowing that the complexities of most things upon which we rely daily are beyond our capacity of knowing, but are nevertheless, as experience tells, reliable and deserving of our confidence. This faith in the reality of all things that make up our very being, our origins and our destinies, is no less certain and reliable than our rationality, that is, than things known to us. In fact, the reality of intuitional confidence, based on past experience, is itself true and authentic knowledge no less credible and deserving than knowledge based on personal experience and communication.

Faith in the unknown and rational attempts to express it, constitute relational (religious) phenomena, namely, statements of belief (creeds) and ritual celebrations (art, liturgies, aesthetics, song, etc.) that glory in the universe of the unknown; together, beliefs and their affirmation in the celebration of relationships embody the essence of religion.

Faith is a social virtue and a faculty of social virtue. From the experience of mutually reinforcing relationships, consciousness learns to reciprocate with good faith responses. Faith consciousness is dually informed, namely, in experience and in knowledge. In experiencing well-being, and in understanding the reasons for well-being, namely, interdependent relationships that originate and sustain all life, self-conscious life grows in the self-assured sense of the reliability of the network agencies involved in personal and social well-being.

In the personal understanding of what *reliable* relationships mean, each is enabled to enter consciously into *authentic* relationships with others, namely, in relationships that are self and other reassuring, and faith instilling.

Faith pertains to essential grounding, which is the centering agency of cosmic community. Faith is the trustful consciousness of emotional rationality, of sustaining and sustainable connections, which, in molecular structure and family, center on femininity, atomic and genetic grounding. We learn to be trustful because we ourselves have been (are) the beneficiaries of the good faith generosity of others. Faith, trust, is breached when pretenses, ulterior motives and hidden agenda, intervene interpersonal relationships.

Since we are communally dependent, the greater depths of valuation are often more commonly experienced in communal celebrations of religious creeds and rituals; however, if the bases of our beliefs, rituals, and celebrations are misinformed, they may eventually have negative effects on interpersonal relationships and personal self-perception. This is why reason and faith must constantly communicate and update, namely, so as to affirm and reprove (authenticate) the creeds and rituals of faith that inform the commonweal.

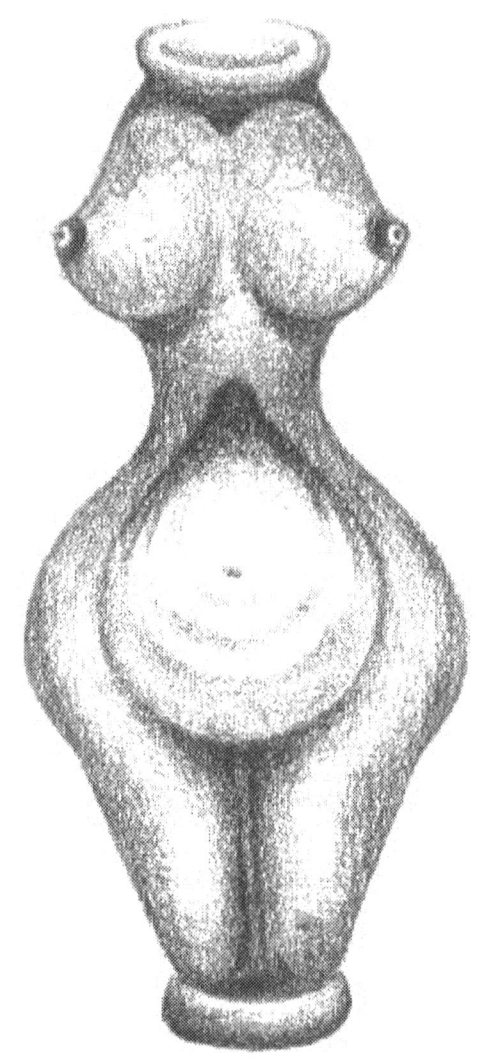

WOMAN WELLNESS
Pencil sketch by Monica R. Steffen
c. 1991

EARTHEN VESSEL

Mystery and miracle,
Mother's words are memory's milk.
God approvingly watched Jesus and John,
Splashing in the Jordan,
Testing the water, and their mothers.

The games grown-up boys play
Wreck nations and bury civilizations
In stratified rubble—the stuff of rebirth, waiting
For baptismal water and the coming of light.

Easy as water from Earthen Vessel
God accedes to woman
Expecting wine from water.
At the sight of the Samaritan Woman's Vessel,
Fresh-filled and overflowing,
Thirsty Jesus remembers and asks for a drink,
Not so much to receive as to give.

Sylvester L. Steffen

The Court of Conscience

The purpose of consciousness is truth. Truth in broad perspective is that which serves the integrity of personal and common well-being. To a greater and lesser extent, every time we make a decision we "hold court", that is, we engage the process of rationality in our coming to a decision and to a conclusion of truth. It might also be said that the purpose of history is also the pursuit of truth; however, truth is by its nature complex and difficult to nail down with precision because consciousness is on a climbing road of complexity whose perspective on truth is also a changing complexity.

From the Christian perspective it seems that history is divided into three ages of consciousness. These ages are: 1.) The Age of Patriarchy, the era prior to the coming of Jesus, 2.) The Age of Christ, the some 33 years of Jesus life on Earth, and 3.) The Age of the Holy Spirit, all time since Jesus' physical life on Earth. But because of the human penchant to institutionalize everything, the Age of Patriarchy is still with us and the Age of Jesus is still with us because of fixations on his physical person. The Age of the Holy Spirit has hardly begun because institutional Christianity in practice scarcely admits, let alone recognizes, the primacy of personal conscience in Christian living and the presence and authenticity of the Holy Spirit working in individual consciences.

The state of the "process of conscience" for lay Catholics is yet unexpressed. Somewhere along the way, before it even gained recognition, it got lost in the logic of institutional primacy, which politically subordinated the primacy of institutional over personal conscience. What institutions do is rationalize their own logic and lock in ideas of their own choosing. These become institutional fixtures that are used as presumptions of institutional preference as it evolves through history. These preferences become fixed, myopic presumptions, whose blind spots become exposed in the course of history. Institutions are reluctant to question their preferred logic and even more reluctant to admit their mistakes. Sometimes the only way change will come is if people abandon the institution until it moves off its ignorance and arrogance. This perversion is general, and it includes Churches and even the judicial system of justice. Their perversions cause institutions to fall into public disrespect.

The formal "process of law" is (should be) a process of rationality in quest of truth. The judicial process begins, after an action is joined, with discovery, the taking of depositions and the filing of interrogatories. The purpose of these is to collect all information that pertains to enlightening the truth as to facts surrounding matters at issue. After factually correct information has been gathered (by lawyers) on both sides of the joined issue the information is evaluated by both parties, in the light of the law and precedence, and arguments are organized and presented to the Court.

The intention and objective of the Court is *to come to knowledge of the truth* and to render an honest judgment based on "truthful" information. Lawyers, as Officers of the Court, are professionally obliged under penalty of law, to serve the truth-objective of the Court, that is, to do nothing that would prevent the Court from rendering a decision that is based on the "the truth, the whole truth and nothing but the truth". The wrongdoing of perverting truth is nothing less than the obstruction of justice. Witnesses are sworn to *tell the truth, the whole truth, and nothing but the truth.* Truth is often many-faceted and because of this it is sometimes conflicted in the perspective of the litigants. The whole point of the judicial process is to clarify conflicted truth—in theory at least.

After the Court is fully informed in the issues it is ready to give its verdict. From its procedures it is clear that the process of law is a formal application of the *process of reason*, which engages the processes of communication and consciousness in coming to a decision. The obligation of the Court is to be faithful to truth, that is, to render a verdict that is honest with the truth, which is to say, "conscionable". The Court functions as conscience on behalf of the litigants who expect the Court to function with integrity.

The process of law and the process of rationality are both "processes of truth". When consciousness is informed with knowledge of all pertinent information (through communication) it is enabled to make a call of conscience, a truth-based decision. All this fine rhetoric goes down the drain when lawyers play games and use the Court for purposes of self-advantage. Faith in the Court is destroyed by the behavior of Court officials who distort truth for money advantages, whether for themselves and/or their clients.

And so it is with us in the use of the process of rationality. If as a matter of habit we are careless with the truth and allow self-interest to motivate all our decisions, then we lose the capacity to function with intellectual honesty. And so it is with people who are professionally engaged in pursuits, which wrongly presume on truth, for example, institutional professionals, whether of church or state.

The question before the Court in every action is: "Whose truth will prevail?" The Court functions as a war arena, not where combat is physical and violent, but where it is civil. Civility and truthfulness are expected of all participants.

How is the prosecution of truth consistent with the Christian advocacy of love? The Spirit of truth is the same Spirit of love. When the perspective of truth is conflicted, the vision of love is obscured. When one feels aggrieved in his truth, love is the victim. The Old Law of violence, an eye-for-an-eye, a tooth-for-a-tooth justice, that is, the tit-for-tat reciprocation of violence, is to be replaced, in the Christian view of love and justice, with a system that gives love a chance to prevail between parties rather than the violent acceleration of recriminations.

The Middle East, the land of Jesus' birth, the announcement place of Jesus' call for a new phylum of love, was already in his time a place torn by war and violence. The violence of tribe against tribe and nation against nation makes the already meager resources all the more dear. Violence and recriminations waste people and natural resources. The people and the land are the sorry victims. Cultured intolerance and exploitation make everybody losers.

Jesus promised that the Paraclete (advocate), the councilor of love and truth would remain with his followers for all time and would be ever active in prosecuting the truth that was given him by his Father, who is in heaven. This Spirit seeks for God's children, all his children, love, not hatred, hurt and violence. As children of the same Father, whether called God or Allah, we are sons and daughters who are inspirited with the same Soul of love and justice—a Soul that advocates for love in the prosecution of truth.

The words and living of Jesus are illuminated in the light of truth. The zeal of the same Spirit fires love in all hearts and illumines truth in all consciences. On the road to Emmaus, the ardor of Jesus' truth kindled in the hearts of his followers as they discoursed with the happenstance comer whom they did not recognize was Jesus. Do we in our confrontations with strangers fail to recognize the presence of Jesus in persons masked by hatred in their hearts and anger in their actions? We too are blind to their love potential.

The terror of ancient violence has visited American shores. Intentional violence is always despicable, but especially when it is done in God's name and under the aegis of God's Church. Violence is a human proclivity. Virtually all institutionalized religions are guilty of prosecuting violence for self-serving purposes—even the Apostle Peter himself and the successors to his primacy. Can the wisdom of Christian love and justice prevail in the world today? Or will we reciprocate with greater violence for violence experienced and aggravate the wreck of the spiritual? Or can we rise above brute instincts and institutional habit and prosecute justice in the Spirit of love, and prosecute against violence by engaging the process of truth, the process of reason, the process of love? Commitment to "trimorphic resonance", the pattern of rationality and Trinity, is the wave of the future, not mutual wars of violent annihilation and alienation. The engagement of the process of reason is nothing less than the dialectic of love.

Jesus witnessed publicly his life's testimony before the heavenly Court, so that for people of all time, his life would exemplify how one should conduct his/her life personally. Like Pilate we can cynically brush truth aside with the easy comment "what is truth?" Unless we have completely suppressed the inner light of conscience, we will experience it reacting within when we are about to do something that is less than truthful. Truth is a struggle of consciousness. Consciousness is a spirituality that advances by its own dynamic. Consistently, consciousness seeks the advantage of that in which it resides, so that as time advances and as new contingencies are experienced consciousness adjusts to the insights of experience. The venues of consciousness are many and varied nevertheless; spirit seeks commonly to serve the nature and purpose of its every venue.

There is one spirit but many witnesses. In Christian consciousness the witness of Jesus is the same Holy Spirit whose wisdom breathed consciousness into the least and greatest cosmic structures. Because the Spirit of Truth is presently active in the process of truth we dare to trust divine integrity in guiding us in our personal pursuits of truth. Our personal humanity is in these times in communication with divinity no less now than at any other time so that the word/work of truth is alive and well in the authentic word/work of every person.

The authenticity of word/work is the Christic authentication of every person for Jesus promised the presence of the Spirit for all time. Since all are inspired in the same Spirit we may truthfully realize that we personally embody Christ Present, who was truth's advocate and who is still in our persons. The Jesus of history is the cosmic Christ, the Advocate of truth, who incorporates humankind in a community of cosmic religion whose work is inspired by the rising word/work of consciousness. Personal conscience requires of each person the task of doing theology, doing truth. Consistent with common rationality each person must do his/her own truth.

Truth allows no easy escape. No one can live blithely under another's truth and be satisfied not to challenge its integrity. Belief that is habituated not in reason but in blind acquiescence in another, whether person or institution, is not faith, it is a deception akin to idolatry for it is attributed to humans rather than to God. Faith lacks integrity except it is founded in one's own personal rationality. Anyone who appropriates the focus of another's faith wrongfully takes the place of God and sets him/herself up in God's stead.

Godlikeness
Nurtured in Nature

GODLIKENESS IS BIOREGIONALLY EXPRESSIVE. EACH BIOREGION HAS ITS OWN DOMAIN OF RATIONALITY (ECOSPIRITUALITY), WHICH ABIDES IN ALL NETWORK MEMBERS. GODSPEAK IS INDIVIDUAL AND COMMUNAL, EVEN AS ARE THE BIOREGIONS THEMSELVES.

THE REGION-SPECIFIC "LANGUAGE" MUST BE SPOKEN BY ALL WHO ARE BORN INTO IT. TO BETRAY THE HOUSEHOLD (ECOLOGICAL) HERITAGE IS TO SIN AGAINST ITS ESSENTIAL SPIRITUALITY, A SIN THAT IS UNFORGIVABLE AND WHICH IN ITS CONSEQUENCE PRODUCES POISONED FRUIT.

As the equation of Special Relativity says, matter is equivalent to energy and is reducible to energy, so, the embodied human is spiritually reducible, *if spirituality is energy-equivalent, if consciousness is energy-equivalent.* As matter presupposes energy (as body presupposes soul), so human materiality presupposes prior (divine) spirituality. In this existential presupposition the *hypostatic union* of the human/divine—its *spiritual essence*—is "from above" while its material qualification is "from below". The *above* (heavenly) and the *below* (earthly) mutually engage each other. (*"Above"* and *"below"* are terms of convenience that facilitate understanding, but, if they are taken to be mutually independent, the terms may defeat in consciousness the unity sense of spirituality and materiality and may cause the misguided elevation of one over the other.)

Spiritual insight (male), perceived as "coming down from above", and material substance (female), perceived as "rising up from below", suggests a separation, a real polarity, a compromised divide between the divine and the human. The conflict of this perceived polarity confuses understanding of the historical, human Jesus and his transcendent divinity, that is, how they work together everyday in his life.

God is the overarching Spiritual Essence *in which all being is.* This "in-ness" of being is the *hypostatic union* of spirituality/materiality. While all materiality is essentially homogeneous in the substance of its composition, it varies in the degree of complexity so that its *spirituality* also varies in its complexity. The capacity of self-knowing is a spiritual complexity upon whose expression depends God's own. Thus, the self-reflective human is in a sense "essential" to God in the *matter* (materiality) of self-knowing goodness. In love's commitment to doing good, human beings are good and are expressive of God's goodness. On being incarnated into human dependency God self-identifies with human dependency, and in the person of Jesus models the meaning of Godlike giving.

From personal experience we know that the faith we possess was made to flourish because of the trustworthy communication of parents. Through communication and experience we discover trust, whom we can trust and whom we cannot. The virtue that we learn from trustworthy communication is faith. As a novice to religious life I recall the words of the novice master affirming that a newborn learns faith as she nurses her mother's breast. Our first confidence was built upon the trustworthiness of our parents. As creatures of natural experience, the confidence of first experience is more immediately "objectified"; but, with age and wisdom, we ascend by the process of rebirth "in spirit and truth", to a more "primordial" consciousness. The institutionally advanced perception of revelation "from above" seems not to be consistent with the personal experience of consciousness rising "from below" in the rational process of rebirth *in spirit and truth*. Institutional theology *from above* puts institutional structure/insight over the personal, and imposes its ancient train of speculative doctrine on the individual even though certain ones of them have lost their credibility in new consciousness. Thus the divisions and alienation of theology *from above* and theology *from below* are sharpened by conflicts between personal conscience and institutional doctrine. Upon faith's ground we develop a self-confident, vibrant expectation (hope) of growing spiritually into self-fulfillment. On the certitude of hopeful expectation we come to recognize within our selves a competence to know and to do what is right. In our conscious pursuit of personal and communal well-being, we develop an intentional motivation of purpose, purpose that intends mutual well-being. The intention to know and do what is good is the willful intention of *love*, and, the willful motivation to act on the knowledge of "the good" is *conscience*. In the *trimorphic process of rationality* (communication, consciousness and conscience), the *trinity family* functions in reflection of Trinity, in recognition of itself as the building block (paradigm) of the larger communal family.

Salvation depends on the liberating power of personal conscience. The liberating power of conscience is most effectively salvific, not just by virtue of individuality but most certainly by the resonance of social harmony. The culture of faith, hope and love is quintessentially communal through the intentional dynamic of communication, consciousness and conscience, even as social emancipation and cultural harmony are natural outcomes of harmonic dialog. Salvation is understood and effected in terms of communal harmony directed toward intentions of conscience, informed in conformity with Godlikeness. Individual consciousness and social are reconciled only through dialog. Individualism and institutionalism, which presume a superiority of God-election, of chosenism, are both wrong even as is the theology of a presumed superiority of the *above* over the *below.*

Salvation isn't some acquisition of spiritual status it is continual work and growth. As process of word/work, salvation is the experience of *sacrament,* the transformational growth of conscious experience in God-Present. Thus, it may be said that salvation is revelation, the product and process of human/divine rationality. Like the expression of Big Bang expansion, transformation into fullness is an expansion process even as are consciousness and conscience. Salvation is the continuing resurrection and ascension of consciousness toward Godlikeness.

Salvation is more than belief it is a conscious endeavor, an experiential intuition of substantiated communion with God. Not to grow is not to become whole, is not to experience salvation. Salvation is mediated in the process of rationality. In the wholeness of cosmic rationality the consciousness of God's presence is a continuing revelation. In the ongoing process of *trimorphic rationality* the "word" becomes efficacious in work, that is, in actions that are required by conscience.

The process of theological virtue, faith/hope/love, the culturing of relationship in social community, patterned after relationship in divine community, is the outcome and process of communication, consciousness and conscience. This "from below" approach to theological understanding, to divine connection, leads us naturally into the real world of "process divinity" at work in creation. In this process of cosmic rationality we discover in ourselves a process of consciousness that introduces us into divine consciousness, which isn't merely "logos" (word) but which is spiritually driven to Godlike action. In exercising the process of rationality we become more and more empowered by the Spirit to collaborate in Godhead Work. In the natural and authentic unity of word/work we experience the "sacrament" of reality.

If Christ is understood as God's presence in the universe, then salvation itself must be understood in terms of universal understanding. The rationality of the cosmos cannot conflict with the rationality of salvation rather salvation coincides with cosmic rationality in which it is situated. Universal "truth" is the authentic consciousness of cosmic rationality, not some "narrow sectarian outlook" (Roger Haight, *Jesus Symbol of God*, 2000, Orbis Books, Maryknoll, NY 10545, pp 188, 189) that claims to be universal truth. Personal salvation occurs in Earth-life experience, for experience underlies the structure of Christology.

Who is Jesus? Every person is born into the divine expectation of being/becoming Jesus-like. All who live in Jesus' awareness of human/divine person are heirs of Godhead standing even with Jesus. Consciousness is possession. Consciousness is effective in that about which it is conscious. Consciousness of good enables one to be good. Goodness is limited to the extent that consciousness is limited. Grace is a facilitator of goodness; to the extent that we understand what facilitates goodness, we understand grace. Thus, the harmonic reconciliation of self with God is less attuned in one less conscious and less consciously motivated. Goodness consists in doing good; thus one who is habituated in doing good is habituated in being good. In our time, a prime example of this reality is Mother Theresa of Calcutta. Her theology is the high perfection of doing/being good. Goodness in all relationships is that which expresses Godlikeness.

In theological language the word "hypostatic" is used to explain the relationship between the human and the divine nature of Jesus Christ. The study of the human/divine personality of Jesus is called *Christology*, while the study of his mission and lifework is called *Soteriology* (salvation message). [*Hypostasis* is a combination of two Greek words, which in English simply mean "understanding".]

Christology, in its radical sense of being/becoming a "new phylum of love" (Chardin), sees God present in all creation, and especially present in every human. Every person is to grow into Christ-identity. Every newborn is a Christ of *second coming,* whose self-fulfillment advances the communal whole-making one step further into greater Godlikeness. In God there is no distinction of male or female, master or slave, only *all-ness.* All are called to the salvation work of "whole-making" in Godlikeness. In God there is no *stasis* in fixated being, for all is *becoming,* coming into fullness of being. Thus, the human/divine concern is rather the *ratio entis* (the reason of being) than the *ens rationis* (the "being" of reason—*the mental object*—figment), for reason is process, not *stasis.* Fixation in *being* is in fact the frustration of *becoming.*

The *hypostasis* of the ascendancy of consciousness is the resurrection of the flesh in recurrent birth, even as the *hypostasis* of the resurrection of the flesh is the ascendancy of consciousness. There is no *understanding* of one without *understanding* of the other. This proposition states the Sacrament of Hypostatic Union. The unity meaning of Christology and Soteriology is Godlikeness. Godlikeness is the great mystery of the Word-Made-Flesh, of humanity in divinity, of divinity in humanity. Godlikeness is the revelation of God, the consciousness of *salvation*—a birthright potential.

Relationship like rationality is a process. Process presupposes sequential, interdependent steps, *understandings* that lead to ever developing outcomes. Personal rationality is the whole complex of understanding that is embedded in the consciousness of one's neural, mind/body complexity. The process of consciousness (rationality) may be described as a three-step (*trimorphic*) process. The steps of the process are continually repetitive in that the completion of one cycle, upon coming to some new conclusion or understanding, becomes a basis for subsequent process-cycles—communication—consciousness—conscience, and on and on. Of course all aspects of the three-steps may be variously happening at the same time. It's not like the steps are clearly distinct one from the other, or that only one step can be happening at a time.

Communication is the beginning step of understanding. If we want to learn about a whole new subject matter, e.g., quantum science, we can start by talking to people who know something about it, or we can get a book or videocassette. Little by little, through cumulative information we begin to construct in our consciousness a whole complex of associated ideas that form a neural web of understandings about quantum science. New learning becomes a new knowledge base, which can access other knowledge complexes already embedded in our neural wiring/storage network.

The process of rationality is the essential, energetic basis of cosmic continuity, which is presently the energetic and conscious complex at work in the diverse webs of network life and its contingent resources. Nature is itself a multi-faceted complex of interdependent entities and contingencies, all of which compose the essential continuity of cosmic evolution (transformation).

The continuity whole of nature is driven diversely by energetic nuances of inherent *spirituality. Transformation* is a process of nature's inherent spirituality, which substantiates all materiality, and which gives to the cosmos its essential continuity. Creation is inseparable in process from the purposefulness of its continuity and from the ever-functioning interdependence of creative Word/Work in process.

The irreligious misdirection of human willfulness intervenes the continuity patterns of Word/Work and brings on humans their own undoing. This is the crisis of conscience that confronts today's global populations. All are called to retrace for themselves a way out of the labyrinth in which all are imperiled. Process rationality remains the true and tested avenue available to conscious life for finding conscionable correctives to its habituated irreligion. Living in alienation from one another, we perish, but joined in community, in communication we survive, we come to salvation.

Revelation and Salvation are the word/work of Godlikeness,
A lifetime of conscious growth into wisdom, age and grace.

Revelation and Salvation
(The Lord is my Light and Salvation)

The matter of *divine revelation*, the way in which God-consciousness impresses itself in reflective human consciousness, is an ancient and ever new occupation of mind. Revelation pertains to the manifestation of divine purpose (intelligence and will), and to divine extravagance in the transformational universe; revelation is an insight into the *divine mind at <u>work</u>* in self-reflective consciousness—"work" meaning the substantive evidence of divine intelligence, and will and "word" meaning the *logic* behind work. Surely, the continuous expansion and amplification of divine presence, also in reflective consciousness, must be no less surprising, varied and manifest than the diversity of flora and fauna on the face of Earth. And, just as these flourish in beauty, ever more grand and intricate, so must the revelations of the divine in consciousness.

It is the fixed position of institutional Catholicism that direct divine revelation ended in Apostolic times; this fixed belief raises the ever pertinent question "How does God communicate divine insights to people of the present time? And, How does divine communication differ now from before apostolic times?" Certainly, "inspiration" has been and continues to be a spiritual activity of human consciousness. In the continuity of consciousness there are spiritual and material components. The continuum of spirituality and materiality is the unified agency and venue of inspiration.

In Christian belief the apex event of conscious, divine revelation occurred in the mind of Jesus Christ, who represents for Christians the divine/human personification of the God/man. Christian belief, as brought forward in the Roman Catholic Tradition, sees the revelation of the divine in Jesus as being so complete as to lead to the conclusion that *direct divine revelation* ended in the times of the Apostles who personally witnessed the life and teachings of Jesus Christ. Roman Catholic belief, while affirming the unique revelation of divine consciousness in the Person of Jesus, also affirms the continuing revelation of the divine by the Holy Spirit in the consciousness of every age and time. This continuous processing of spiritual newness is a baptism of Spirit in "consciousness and truth", whose fullness becomes manifest in the consciousness of the times. In its evolution over time, consciousness becomes ever more subtle even as reason becomes increasingly informed in the infinite varieties by which the Spirit is revealed in the word/work of cosmic creation.

Authentic living is a life-quest. It is discovered in the dialog of faith and reason. Faith is inspiration *from above (within, transcendent)* and reason is inspiration *from below (senses, immanent)*. Revelation isn't a matter of the speculative elaboration of fixated presumptions; rather, it is a matter of insight occurring within the framework of interpersonal dialog, emerging from contexts of historical consciousness. Religious truth is secured in the context of the authentic emergence of consciousness.

Human will and intellect (faculties of experiential intelligence) are subtleties of quantum-electric consciousness. Consciousness, like quantum-electric potential, is both reflex (in-tensional) and reflective (intentional). Consciousness is in part *finite*, that is, it terminates with individual death, and is in part *infinite* (transcendent), that is, it is communicable by word, and genetically by the inheritance of genetic imprints. [Persistently nurtured elements of culture become psychic imprints in genetic consciousness, *packets* of instructional imprint, social codes, mores, etc., in Dawkin's word, "memes".]. Thus, intelligence and will, as aspects of consciousness may be codified and become "transcendent", i.e., handed down from generation to generation, in physical and psychical codes. It is from the evidence of human experience that we can conclude that intellect and will, like consciousness, are partly finite and partly infinite. Genetic coding and its effects are experientially driven.

Another word for experiential consciousness is "rationality", *cosmic* rationality, for all consciousness originates in the evolutionary continuity of cosmic experience. The highest intentional motivation of purposeful consciousness is love, which is manifest in the evolving continuity of cosmic word/work; the experience of grace (spirituality) in working matter is called "sacrament".

Sacrament has before been characterized as *sign* (material and visible) and *grace* (spiritual and effective). A way perhaps to clarify is to say that *sign is work* (nature) and *grace is word* (nurture). The soul of Sacrament is the consciousness (Mind) that empowers Word/work; the empowering mind of word is intellect and the empowering mind of work is will; together intellect and will are the empowering agencies of natural sacrament in its *intentional mind*, that is, in its *ex opere operantis* agency. Intentional (reflective) agency is the agency of self-aware purpose, whereas, the *reflex* agency is the unreflective agency of *natural intension*, the *ex opere operato* agency. Both agencies are effective in purposeful outcomes.

Trinity, the spiritual essence of creative agency, operates at many natural levels, e.g., intellect and will, intension and intention, grace and sign, word and work, femininity and masculinity, soul and substance, nurture and nature, quantum and electric. In their joined doing, paired agencies effectively process a third *mindful* outcome of the same essence as their combined essences. The communion and community of Godhead Agency are everywhere revealed in the soul/substance of creation working. The Scripture of God's Word is everywhere and always being "written" on the parchment work of the cosmos. Because God inheres all things, human beings, even like Jesus, also reveal the divine sacrament of hypostatic union in their communion and community. By their love for one another, the purpose of cosmic inherency is revealed.

Love is the spiritually refined fire of the Big Bang, the universal inspiration of all intentional/intensional word/work. Love is rationally refined spirituality. As such, it manifests the Spiritual Agency inhering (panentheism) all reality, whose Providence we identify with God. God operates in love's rationality and is revealed in nature's holy purpose. Worship consists of living faithfully the purposeful codes of common well-being. Individuality originates in the purposeful codes of well-being, thrives and subsists for a time subject to them, but must return to the greater purposes of transcendency, to the spiritual advancement of communal well-being.

To the primitive mind, the revelation of the divine seems more revealed in some minds than in others. People with a greater sense of natural insight were revered as specially informed (inspired) by God. The shamans, prophets and priests of ancient times were perhaps more intuitive in things divine and mundane than most of the population, and who were then sought out for their insight and guidance in matters divine.

In view of this understanding, it is appropriate to expect that theology does inform people authentically in the circumstances of their times, and, to ask yet again, "How does theology, updated, inform God-consciousness in the contemporary world?" The answer that seems relevant to every time is that human transformation in divine consciousness occurs along with the transformation of cosmic consciousness, reflectively and collectively present in rationality; both of which, consciousness and rationality, advance by communication.

The word used for verbal communication is *dialogue*, the exchange of ideas by back and forth conversation. [*Dialectic* is a means by which differences can be rationally reconciled, for all consciousness occurs in the universal venue of cosmic rationality.] So, in bringing the rationality of word to a statement of common relevance, we may say that the ongoing dialogue between "scientia" (knowledge, reason) and conscience, "*con*scientia" (informed judgment), is a way in which self-interest and common interests can be discovered, reconciled and authenticated in relationships motivated by the theological virtues of faith, hope and love.

As a process and product of "whole-making", (salvum facere), salvation is a process of revelation and of reconciliation. Salvation is more than mere belief it is a sense of real resonance, of experiential harmony (union) with God. Salvation is a word/work process of growing into the wholeness of being/becoming. Fullness of consciousness is organic for we grow into God's presence as we grow in wisdom, age and grace. Salvation is mediated in the process of rationality. In the wholeness of cosmic rationality the consciousness of God's presence is revealed. In the ongoing process of rationality the word becomes efficacious in work, that is, in action required by conscience. Salvation isn't a location, a onetime event, an endpoint to be attained, or an endgame to be won; rather it is the continuing (moment-by-moment) resurrection and ascendance of consciousness. Not to grow in consciousness is not to become whole, is not to experience salvation.

Coming to salvation is more than merely reaching a spiritual or physical status; salvation is work and growth. Spiritual aspiration—

salvation/redemption—is 10% inspiration and 90% perspiration. As process of word/work, salvation is the experience of sacrament, the transformational growth of conscious experience in God-present. Thus, it may be said that salvation is revelation, is redemption, the product and process of simultaneous human/divine rationality. Salvation is an outcome of focused intention to know and always do what conscience discerns to be right. Salvation is a labor of love; revelation is love's insight; and redemption is self-correction in revelation's light—the continuing inspiration of reason that uplifts consciousness. Love is life's past, present and future—the discovery and motivation of conscious ascendancy ever in process.

Love is expressed and experienced in relationships of mutuality, of symbiotic intention. Love's amplification sounds in man/woman mutuality. In the mutuality of the sexes natural ambiguity and openness prevail; a spiritual component of which is perplexity, uncertainty and confusion. These are natural to sexuality because of its creative openness. The dialog between *instinctual* sexuality and *rational* sexuality is loaded with tension even as is the intercommunication between molecularly *conscious* intension and intention. Tension (structural friction) resolution is the task of symbiosis, mutuality. The consideration of mutuality obliges every interpersonal involvement, whether same sex or heterosexual. [Mutuality means *reciprocal benefit*, in which both (all) parties experience real advantage from the relationship, not harm.]

Cultural patriarchy, religiously advanced, has fathered the social anomaly called "sexism". At this time, Roman Catholicism is confronted by a great crisis, the crisis of the publicly exposed sexual abuse of children by priests and the conspiratorial cover-up by the Church hierarchy—a problem associated with sexism. Sexism roots in and is advanced by ignorance, arrogance, alienation, elitism and social dysfunctionality. All of these violate male/female mutuality, communion, community and vital continuity; the sacrilege of these desecrates the processes of familial and societal harmony.

Sexism is hyped male egoism (and/or female). Hyper extenuated egoism suppresses familial trinity, community and civility. Egoism is at the root of idolatrous sexism and institutionally frustrated sacrament. Sexism fundamentally denies God's pronouncement of goodness in all creation, and it desecrates male/female mutuality. Sexism is a black hole drain of love's energy.

Egoism (sexism) is an agency of incivility. In the words of Edward Shils, "Egoism is not conducive to civility; the dissolution of society into an aggregation of demanding individualities is not conducive to civility. (Pg 10)…Civility and ideological radicalism are irreconcilable". ("The Virtue of Civility", edited by Steven Grosby, Pg 14, 1997, Liberty Fund, Inc., 8335 Allison Pointe Trail, Suite 300, Indianapolis, IN 4625-1684) Because the energy of sexism is socially dissonant it wreaks havoc on communal relationships.

Male pretense over females is an especially pernicious sin for it presumes divine electionism. Brazen male egoism projects itself into the Godhead persona while it excludes women from identification therein. In the evolutionary history of life, brute male instinct has come to be habituated in the compulsive sexual domination of males over females. The sexual compulsion to dominate is hard wired in the neural patterns of the avian/reptilian brain; however, in *homo sapiens* the consciousness-processing of the rational cortical brain is empowered to override the hard wiring of raw instincts. Because habituated repetition comes to be imprinted over time, social culture can push compulsive sexuality in either direction; it can push sexual compulsion in the direction of brute domination or it can soften compulsion by the rationality of higher intelligence. Cultural domination by males, when religiously sanctioned, invites "in God's name" all manner of personally and socially pretentious behavior. Male celibacy for example, sanctioned as a precondition for membership in institutional religion's hegemony, is an endorsement of male elitism that inappropriately presumes on and self-assumes divine privilege. The Roman Catholic priesthood has come to be weighted with elitist presumptions that violate the personalities of priests as well as of those subjected to priestly arrogation. After the male priestly example, husbands justify their lording behavior over their wives and children.

Priests are trained to think of themselves as "married" to the Church. [As to their personal expression of sexuality, priests (and vowed religious?) are, in John Paul II's admonition, to be "eunuchs" in the Church's service.] In either event, sexual commitment (suppression) is a pillar to the rationale of the celibate and exclusively male priesthood. Does not the fact of male exclusivity make clerics, priests, *de facto*, by definition, "homosexual", that is, *same sex*? In their priestly training, students to the priesthood were/are required to minimize and avoid interactivities with females for such activities may "tempt" the student to opt the heterosexual preference rather than the clerical. (Michael Papesh, "On the Demise of Clerical Culture", AMERICA, A Jesuit Magazine, Vol. 186 No. 16, Whole No. 4569, pp 7-11, May 13, 2002). For the priest, the Church is his female partner for life. Casting the Church in the female characterization ("bride" of the priest) is problematic for it overlooks the simple fact that males also make up its membership. The female characterization of institutional Church seems inept and abused. Theologically, psychologically and realistically it seems more apt to have the priesthood of the Church reflect its ambivalent (female/male) membership. Damaging theological fictions evolve from the exaggerated bride (Church) groom (priest) metaphor. In it the sexual persona of the priest seems not to find adequate expression; in fact, the too-far extension of the metaphor might be personally and socially damaging for it may feed male consciousness with an overload of narcissistic fantasy and sexual distortion with decidedly abusive overtones.

It needs to be said that the freely elected option of celibate living of dedicated men and women in the service of fellowman is an extraordinary grace of self-sacrifice. The institutional requirement of celibacy as a precondition for priesthood is an impediment to the universal obligation of priestly service that obliges everyone in the divine order of communal ordination and does a disservice to the Church.

From the sin of sexism, humankind is in need of salvation. Salvation from sexism might begin with the authentic recognition of sexuality as the necessary basis of evolutionary life. Sex is essentially also a consciousness of mutuality, of fundamental co-dependency. Sex is the elemental "two-ness" of life; in word and work, sex is the trinitarian sacrament of transformational vitality.

Maleness is meaningless except in communion with femaleness. In conscious mutuality reside communal redemption, hope and regard (love, worship) for the divine Pattern of Trinitarian Sacrament. Male pretension violates Trinity; whether in word (consciousness) or in work (communal edification); violated sacrament is a desecration of divinity. Sexism is the cultured schism of conscious and radical mutuality, the exploitation of the *radical other* (femaleness) for the passionate self-advantage of male obsession, possession, and exploitation. The idolatrous egoism of hierarchical patriarchy violates authentic (cosmic) consciousness, mutuality and communal harmony. The societal wreck of basic civility testifies to historically falsified religion and to idolatrous worship at the altar of male egoism.

Roger Haight identifies the question of what we need to be saved from as a "religious question". (Ref: "Jesus Symbol of God" pg. 354). The lesser good and bad things we do, in our relationships with God, fellow humans, and with all "Other" upon which we depend in the natural order, have negative effects from which we need to be saved. In the words of Roger Haight:

> "Any and every salvation theory must respond to the negative experiences of *ignorance, sin, guilt, suffering and death* (emphasis edited)...Salvation must be something that can be experienced now...Salvation must be integral; it cannot touch a so-called spiritual dimension of a person's life and not include his or her activity in this world...The world is the full measure of the human body, and one's particular world contributes to the identity of each human being. Salvation must incorporate the world insofar as the world, although in one respect over against the self, is also a part of the self...Salvation must be interpreted not only individually but also socially. The idea of an individual salvation apart from the salvation of the species is incoherent." (Id: Pp 355, 356).

It is in personal/social transformation that we come to revelation, redemption and salvation. From what do we need to be saved? We need to be saved from: 1.) <u>arrogance</u>: every sin has a component of arrogance, of setting self against God in spiritual/material things/judgments; of setting self ahead of neighbor; of valuating personal self-interest above natural necessity; as Scripture says, "before the fall, pride". The recent public airing of sexually abusive behavior in the Roman Catholic priesthood is increasingly sensed as an aggravated deviancy associated with narcissistic male obsession, and with habituated overtones of systemic repression, ignorance and arrogance. It has aspects of instinct (genetic) and culture (memetic). 2.) ignorance: the saying has it that "a little knowledge is dangerous"; in blind disregard we rush in where "angels fear to tread", and in so doing we wreak havoc on natural necessities and trash God's Primary Scripture. Like "bulls in the china closet", males have behaved badly and continue to do so.3.) guilt: from patriarchal guilt; we need to be saved from the culpability we acquire due to our acts of arrogance and ignorance; from the heavy desperation people suffer, nature suffers, due to our knowing and unknowing acts of wrongdoing. 4.) suffering: from the impoverishment others suffer due to our excess regard for self and the over-burdening we inflict on Earth and others by our hyped appetites and obsession for things; from the specific suffering of women and children caused by hyped male self arrogation, whether in presumptions on God or nature. 5.) death: from the mortal injury we inflict, and from the dying others are made to suffer because of our over extended, consumptive habits; from the spiritual death we impose on our own consciousness due to our ignorant and arrogant obsessions; from the desecration of nature; from the trashing of network life and the extinction of species—an abortion that profoundly diminishes life's spiritual/material depth and resource base for unborn and untold generations.

The three-step "process of rationality" is nature's patterned mechanism of social interaction and interpenetration. The sincere act of engaging the *process of reason (conscience),* so as to be informed in the consequences of actions, is an obvious and necessary component in the advancement of personal/social salvation. Also quite obviously, salvation is a process and a process-outcome, and that, while salvation is a gratuity of divine beneficence, it is also an outcome that depends on personal input. As the saying goes, "God helps those who help themselves"; to paraphrase it, "God saves those who save themselves".

In recognition of the obvious, we are compelled to socialize our efforts toward salvation, namely, by communicating with others and by publicly confessing our uncertainties, doubts, aspirations, failings and successes. Through acts of confessional openness our personal consciousness expands and becomes more open, and the reciprocal transparency enlightens conscience in personal and common well-being. Individual neglect and social failing to engage the process of rationality has the bad effects of deafening, deadening and instigating rigor mortis, as opposed to the good effects of faith's communication, hope's consciousness and love's conscience; in the light of The Way, The Truth and The Life, redemption and salvation are outcomes of these good effects—the sign and grace of rising consciousness.

Theology in Process

Theology pertains to knowledge about God. Cosmic Rationality is humankind's conscious and direct link to God. Theology is an evolving, coincident consciousness of natural (cosmic) dialog. As a category of transformational consciousness, theology must change even as consciousness must. Fixated beliefs (fideism) may obstruct change but they cannot eliminate the need for change. It may be argued that Catholic institutional dogmatism obstructs the process of theological evolution in counter-intuitive ways, that is, against the natural and essential process of cosmic rationality.

The Roman Catholic laity and the laity and clergy of denominational Christianity hold in common many faith traditions regarding the person and teaching of Jesus Christ, which traditions are theologically elaborated in concepts and language that are highly specific and fixed in Christian consciousness. If one seriously believes that theological defects of concepts and terminology are embedded in handed-down faith, one does not easily discard hallowed traditions if one would endeavor to expose the defects and expect the Christian faithful to be open to newness and change.

This writer has been reared in traditional (pre-Vatican II) Catholic belief and holds it dear even now, even though he is conscious of fixations that work against the Church's credibility and effectiveness. Faith tradition does not easily acquiesce to change or challenges, especially if they appear to be spurious on their face. If one challenges the faith, but is lacking in understanding of accepted theological concepts and terminology, the challenge is not likely to be taken seriously by clergy or laity. Aware of this reality and of the need for change, this writer believes that "neologisms" are perhaps unavoidable and that existing theological language needs to make room for evolved cosmic rationality as well as for new theological insight arising from it. The ramifications of theological elaboration are labyrinthine, and what validity they have should not be diminished by contributions of new theology that are derived from new cosmological, biological and psychological understandings.

A shift to a starkly contrasting paradigm, from the long held "static, centrist worldview" to the "transformational worldview", for example, challenges the theological language and beliefs that rely on presumptions of ancient cosmological staticism and centrism. Traditional theology, elaborated in reliance on the static-centrist paradigm, is radically confronted by theological considerations that originate from reliance on the transformational paradigm.

The semantic and conceptual conflicts that arise from the irreconcilability of the two paradigms (closed-ness versus openness) will have to be resolved. Civility and the tranquility of civilizations are at stake; on this matter, the papacies of John XXIII and John Paul II seem to be at odds, radically. John XXIII recognized explicitly a need for new theological inclusions due to general acceptance of the "evolutionary" worldview, "The human race has passed from a rather static concept of reality to a more dynamic, evolutionary one. In consequence, there has arisen a new series of problems, a series as important as can be, calling for new efforts of analysis and synthesis." (Joseph Gremillion, The Gospel of Peace & Justice, 1976, Fifth Printing, Mar.1980, Pg 247, **Gaudium et Spes**, Intro. Statement, No. 5, Para 4, Orbis Books, Maryknoll, N.Y. 10545), while John Paul II clings to the pre-Vatican II theological staticisms of language and concepts.

The inclusions of new understandings under old theological terms and the formulation of new theological language designed for introducing new theological concepts require knowledge, sensitivity and considerable skill. It is difficult to introduce "new theology" without seeming to be disrespectful of traditions of expertly elaborated theology, especially, if a non-expert introduces the new theology. Certainly, it seems obvious that ancient Christian concepts and traditionally hallowed terms need to be newly elaborated from the quantum-electric (acentric, transformational) perspective, terms such as, sacrament, trinity, hypostasis, redemption, transubstantiation and salvation. And it shouldn't be surprising that old concepts and words may not be adequate for expressing a theology corresponding to modern cosmological consciousness. In any event, the inclusion of new theological meanings under old terminology and the development of new terms for new theology call for the deft exercise of neology.

The inadequacy of traditional theology to fit changing social circumstances, such as institutional exploitation of people and resources, has created the need to reconsider the theology of "natural" providence and the role of human conscience in sustaining it. This is something that Liberation Theology has been doing. The desperation of indigenous peoples caused by colonial exploitation and aggravated by politically enduring patronage, has created a theological problem that old theology cannot resolve. The simplistic linearity of traditional Christian Theology is neither credible nor effective in meeting the intelligence and needs of present day people. The "Rock of Peter" has been cemented to the politics and philosophy of old, despotic Rome. Roman politics and philosophy violated human rights in the dispensation of patronage and the enslavement of people. Liberation Theology has surfaced most vigorously in the very countries where Roman patronage and exploitation were transplanted, namely, in Central and South America.

If Christianity is to fulfill its mandate of "preferential option for the poor", it can no longer be a party to propping up unjust colonial structures, which continue the old ways of patronage and the exploitation of indigenous peoples. The Second Vatican Council committed Christianity to deconstruct the political and economic structures of colonialism that repress people and exploit resources for the profit of the landed few.

The recidivism of the papacy of Pope John Paul II to the controlling rigidity of pre-Vatican II has come down heavily on Liberation Theologians who have taken up the task of developing a theology suitable to the needs and conditions of Central and South America and other Third World nations. In this necessary work, Liberation Theologians have had to have recourse to neologisms. The task of theological/scientific construction, in the modern day context, seems not only useful but also necessary if the shredding of Earth's garment is to be avoided and the needs of people are to be provided for. If denominational Christianity is to break free from its theological fixations in staticism, centrism and electionism, new theological concepts based on new societal and scientific consciousness will have to be developed and adopted.

In light of the dynamic nature of transformational consciousness, it seems that the exercise of neology is necessarily a continuing process. While institutional theology may find change messy, institutions will become dated and irrelevant if their theology is not kept contemporary. No doubt it is institutionally more comfortable to hold on to old terminology as much as possible for it is surely no easy matter to fit understandings of science and theology within a common language; nevertheless, such common parlance needs to be developed if critically important knowledge is to be absorbed by popular culture and used in the service of communal well being.

Necessity and Revelation
(What makes sense.)

Evolutionary transformation is driven by its own inner rules; and as energy/matter transforms and becomes more complex in their joined qualifications, so do the governing rules. The complex evolution of self-aware consciousness, energetically interwoven in matter's fabric, includes an unfolding awareness of the divine, that is, of the sustainable mentality that endures essentially in the continuity of cosmic purpose. The ladder-climb of vital consciousness is a refined, helix-bound process of complexity that is both ordered and open. The coming together of complex spirituality/materiality in vital substance is an inherent and self-driving process that engages and reveals cosmic potentials contained in the infinite virtues of the divine. Awareness of *the self*'s changing complexity is a kaleidoscopic insight that, when kept open, continues to incorporate new forms, textures and colors in its ever growing mix. The expanding quest of consciousness persistently self-penetrates the consciousness of the divine with ever more dazzling results.

Consciousness takes on structure and energizes structures. Structure is a necessity, but it is also a potential hazard. It becomes a hazard when it attempts to fix a *status quo*. One form of structure is doctrine, which may be the innocuous casting of divinity consciousness in terms that establish, but leave open, the continued working of the divine. Institutionally fixated doctrine may become *dogma*. It is in the nature of institutional logic to cast dogma in terms that suit the time-framed politics of the institution, and thereby to attach meanings that are favorably restrictive and closed to interpretations.

Cosmic *word*, the revelation of the *noumenal* is inseparable from the revelation of the *phenomenal*, the substance of cosmic *work*. This word/work ("*sacrament" reality*, a term of religious art) is the basis of cosmic necessity as well as of the self-revelation of the divine. God, in quantum-expressions, self reveals in human consciousness, as the grand-eloquent complexity of vitality enmeshed in the logic of cosmic necessity. All that is part of this necessity-complex must conform within it or perish. Invention must stay within the open potentials of sustainable complexity. Divine revelation conforms to the conventions of its own inventions. Cosmic necessity can't always find a niche for every scheme, but for *what makes sense* it finds a place. Fidelity to cosmic necessity presumes cosmic rationality—rules by which energy/matter transforms.

Informed insight, conscious conformity to cosmic rationality, leads to conformed behavior, to moral conduct motivated in sustainable self-awareness (love). Conscience is *awareness and action* (word/work) motivated in love. Conscience as the word/work of cosmic rationality shows God's Face in humanity.

The popular conception of "doctrine" might be understood quite differently from the cosmic, quantum-rational perspective. How might we understand doctrine from the cosmic perspective? Doctrines are the rules of the game. What game? The cosmic game of life, whose rules bring fullness, *whole-making* (salvation). When were the rules developed? They were developed in and with the process of the formation of the *stuff* of life. To whom do they apply? They belong and apply to everyone, to cosmic universality, to everything, in ways that are *fundamentally* the same. What is the nature of doctrine? Doctrine is physical/psychical, spiritual/material. What does it do? It controls the processes, the interactivities of the stuff of life: vital structuring of air, water and soil, cell formation, symbioses; it governs the process of "word" *becoming flesh*; it physically empowers the body and psychically empowers spirituality/consciousness. Whom does it benefit? It is of equal benefit to all for it "equally" empowers all energy/matter. [Genetics are volatile in their *openness* to express and show "equality" in different ways according to the "text" (parental contribution) and "context" (exposure to environmental variables) qualifying individual life.] In what way are doctrines *religious*? They are "religious" in the sense that they govern *relationships* that serve common well-being and are spiritually/physically compelling.

Doctrine is perhaps generally understood as the body of institutional church rules and beliefs that require personal obedience and belief. Given institutional loss of credibility, it is hardly surprising that many are now coming to think that doctrine is an institutional control mechanism that has been used to hold people in the service of the institution. People sense that they are cogs in the service of machinery driven by hierarchical engines, which have no interest in what people think other than that they serve to fuel and serve the machinery. Investments in institutional structures, premised on a theology of Vatican I, hold bishops, priests, and religious institutions, hostage to old mindsets and failing structures. Increasing numbers of lay people accept neither the theological nor the structural fixations that presently cement church to the past and close doors to the future. Church openness to and acceptance of lay insights in theological and structural matters might restore credibility to church leadership. The hour is getting late for the infrastructure to change.

Before a quantum leap, an electron-jump from fixations can happen, sufficient potential has to accumulate in the public battery so as to overcome the institutional hyper-charge that presently holds sway over the "faithful". Transformation will come though it yet remains to be defined; it is only a matter of time. A little open mindedness within church institutions would help hasten the day of real church renewal, the day when lay participation functionally happens in the church body, psychically and physically.

Trinity in Nature

The Great Cosmic Mystery is the Self-identification of Trinity in Nature. Cosmic Wisdom reveals that communion/community is embedded in the "purpose" of Trinity, which is to say that Trinity is embedded in Nature and Nature is embedded in Trinity. The Self-realization of this spiritual power is the Great Sacrament of God Present in cosmic soul/substance, in the Dialogic Process of Nature.

The impenetrable Mystery of Trinity, of Reason, is that It is both the Cause and the Consequence of Trimorphic Resonance; the Mystery of Faith is that It is both Cause and consequence of Communication; the Mystery of Hope is that It is both Cause and Consequence of Consciousness; the Mystery of Love is that It is both Cause and Consequence of Conscience; the Mystery of Purpose is that it is both Cause and Consequence of Communion/Community. It is the Godhead Mystery that the Processing of Purpose is the "Alpha" (cause) of Communion and the "Omega" (consequence) of Community. Beginning and End, Agency and Agendum, are intimately one, *purposely* infinite.

God is neither male nor female. Yet the essential "sexuality" of cosmic soul/substance manifests Trinity fundamentally in the mutuality of "two-ness". In spiritual terms, the positive and the negative, the Yin and the Yang, empower the embodiment of atomic/molecular substance, even as the duality of human personhood (female/male) is the communal embodiment of community. In individuality, the human person is soul/body characterized in rationality, fidelity and intentionality.

Reason, faith and purpose substantiate communal civility; without the agency of their interactive inspiration, humans fail in civil, harmonic relationship. Intentional harmonic civility begins in the individual, in the experience/exercise of mutuality in family.

The comparts of the atom, the nucleus and electrons, and by analogy the comparts of the zygote (the ovum impregnated with the sperm), in their own situations, *commune* in twosome union to become a threesome *community;* in the first case the fruit of the union is the atom, and in the second case it is the zygote, the genetic beginning of a child. Sexuality is the cosmically joined intimacy of energy/matter, of embodied energy, fruitful in Trinity. The *polar* comparts of vitality, femaleness and maleness are, of themselves, functionally incompetent, but in their mutuality both become competent. The male rejection of essential grounding in femaleness is an alienating exercise that is fraught with excessive self-assertiveness; undisciplined, it is an agency of insidious havoc.

Whether or not self-aware consciousness is the same as "soul" doesn't matter, for whether it is or is not, soul/consciousness is the spiritual grace of human *being/becoming,* which is an effect, product, result, and consequence of the "procession of trinity" in the natural order of cosmic transformation. Cosmic transformation is a natural power of quantum relativity, which is intuitively evident in human experience.

Quantum science has penetrated the mystery of the cosmos to a degree that the interactive mechanisms of wave/particle are better understood, as is the quantum-electric *intension* of energetic particles, which have the ability of amassing matter in energetically unique qualifications that are greater than the mere sum of qualities possessed in the contributing components. An obvious case in point is the coming together of two hydrogen atoms and one oxygen atom to form the water molecule. The water molecule exemplifies the *trinitarian construct* from two individual entities together transforming to produce an entirely new, third entity—water. Like the water of life's origin, consciousness belongs to the *sea of infinite substance,* whence human consciousness derives its self-aware subtlety.

Religion is quintessentially about relativity (relationship); which means to say that it isn't just about *nature* (genetics), it is especially about *nurture* (memetics). Conscientious nurture is the intentional equivalent of intension in nature. Genetics advance the vital *text* (primary scripture*)* of open being, while memetics qualify the text of being/becoming to the *context*, that is to say, nurture attenuates the harmonic energies of nature while it also accommodates as it must to disharmonic energies. Intention weaves context into text.

Even an elemental understanding of biology recognizes that sex preeminently enables life's evolving diversity and vitality. First and foremost, sex is agent and agendum, cause and effect, of original and ongoing communication. Song, dance, color, artistry are all devices of communication that manifest life's diversity in its varieties and extravagance. The division of humankind into female/male difference is a communication mechanism by which divinity self-manifests in the creativity of natural self-awareness.

In its service to higher purposes, intuitively conscious sex pursues its perfectibility. In serving its own perfectibility, sex has in open evolution burst forth in wildly extravagant diversity and variety. It is in the extravagant expression of species diversity that sex enables the more sustainable traits of vitality to surface, survive, renew and find new and self-perfecting variations. The pathway of genetic testing and perfecting proceeds patiently and is responsive to the day-by-day context in which it operates; the pretentious rushing of the process by humans is foolhardy, idolatrous and fraught with trial-and-error tragedy.

The memetic work of nurture is *symbiosis*. Nature and nurture are cause and effect of each other, and by their mutuality, soul/body subtleties are raised to higher states of psychical/physical being/becoming. This is the "resurrection" edification of the transformational paradigm; this is the Great Mystery of Trinity into which each is born and destined in the divine. The intuitive lesson of cosmic rationality beckons to everyone to be sustainably faithful to nature/nurture's essential rationality. This is the universal call of priesthood, which is also the common moral obligation imposed by birth on each individually, namely, to advance in one's personal word/work the nurtural and natural authenticity of life's *primary scripture*.

The role of excessive self-assertiveness in male-evolved theology is a deception that denigrates femininity, sexuality, and defrauds humankind of an authentic and holistic sense of self in the *image and likeness of God*. Except and until institutional male hierarchies accept the leadership role of women on an equal basis with men, males will continue to advance idolatrous pretenses within their institutions, and disregard divine order and ordination. Universal ordination in priesthood hallows every human person in a "chosenism" that identifies every person *as gift and giver*, divinely sanctioned, to expend oneself and to serve the spiritual, personal and social betterment of self and all by personal commitment to redemptive works, which come to be conscionably clarified in the rising consciousness of cosmic rationality.

All life belongs to the natural order of Trinitarian Processing, an order which, as imaged in Botticelli's Venus rising pristinely conscious from water ready and willing for her vital, transformational role, is motivated interiorly by the continuity purpose *intending* all soul/substance. In her self-aware, personal *intension*, and in her committed *intention* of conforming her whole person to the Trinitarian Processing of human life, woman's divine mission is reverenced in the divine representation of her and in her personal "fiat".

Just as Venus rising from water and Mary's *fiat* affirm the divine role of femininity in soul/substance transformation, so, the "religious" institutional orthodoxies that compel the total concealment of woman in draped garments and that deny her a role in Church and public life, seek to diminish her and alienate her from her divinely ordained role in the cosmic order of Trinitarian Processing. In the traditional male theology of Trinitarian Processing, the role of the male is exclusively incorporated in the Godhead, while in real life his role is less significant than that of the female, as seems to be acknowledged in the Church doctrines of Mary's Immaculate Conception and the Virgin Birth of Jesus by her. These doctrines suggest a changed theological emphasis in which the feminine is accorded a place in Godhead consciousness, in contrast with the prior all-male emphasis of Trinitarian Godhead based on a misinformed culture of belief that the male alone originates the pristine (male) human in his body and that the female merely nurtures and brings to birth the divine male issue. The feminine is the quintessential *sign* and *grace* (work and word) of vital Sacrament, of transubstantial Eucharist, divinely ordained in Nature. In authentic worship of God-Made-Flesh, woman is, like Jesus, "altar and sacrifice". Inextricably, femininity is manifest in Godhead Divinity, in sacramental word/work.

It is only in twosome mutuality that the ambiguous potentials of creative life flower in infinite splendor and diversity. The universally compelling excess of biodiversity is a sign of the cosmic rapture of the deep-sex unveiling of the *Heart of the Rose*, the Mystery of Trinity.

Sylvester L. Steffen

The Heart of the Rose

Our lives are cluttered with fetishes and superstitions. These are included in the baggage we bring forward from our varied pasts. I carry as much useless baggage as anyone. And I must acknowledge that I expound on my obsessions ad nauseam, and I excuse myself, rightly or wrongly, on the belief that if they are not exposed for what they are we will continue to credit them more than they deserve. Useless baggage gets in the way of meaningful living and of being light and inspiration to one another. This baggage is called *being full of self; putting personal well-being above that of others; being preoccupied with stuff, with objects, "quanta", and failing to recognize the "is-ness" constituting things, that is, their inner energy, their potential for good and/or hurtful usage, the hidden "qualities" inherently woven within everything, the quantum-physical power to transform*—the aspect of "physical" that is traditionally referred to as *metaphysics.*

It needs to be acknowledged that the baggage excess we carry is mostly the witless fabrications of male obsession, aggravated also by the systemically cultured distrust of professionals of science, philosophy and theology—each making exclusive claim for turf, for personal and institutional reasons. In reality, *physics* and *metaphysics* deal with one and the same cosmic substance (physics, quantities) and soul (metaphysics, qualities). Splitting the disciplines of inquiry into separate professional realms is a schism of artifice that is loaded potentially with cultured alienation—one might truthfully say "cultured insanity".

We fight change; we fight consciousness when it compels us to change from old habits and ways of thinking. We seek comfort and escape in the world of stuff and thinking into which we were born and which we cultivate. We fail to see inner qualities of things that make them *subjects with qualities* that can bring about changes for the better in our lives and in the lives of others. Nature endows us with powers to make changes that affect our future and the future of life on

112

Earth, for better and for worse. Of course there are things in nature over which we have no control, but, in the way natural "things" are used, we do have very telling control with telling consequences.

Choices, options, come to us day-by-day, instant-by-instant; choices that can maintain the *status quo* of baggage overload, or choices that can move beyond heavy stuff and get to *the heart of the rose*.

Words can be heavy stuff. The professions are full of it, especially, science, philosophy and religion. Religion has even tortured and killed people for the words that they have uttered publicly. So have governments. And they still do. Personally, we pick and choose friends and enemies on the basis of their words. Hot arguments are still generated over words like body and soul, who God is, what belongs to Physics and what belongs to Metaphysics. These arguments go back before the time of Aristotle, whose distinctions of words and thought are still heavy in Roman Catholicism. Words still have the power of dynamite in wrecking human relationships.

Human (cosmic) experience pleads the cause of the oneness of physics and metaphysics, the transcendent oneness of soul and body. We do not experience body without consciousness nor do we experience consciousness without body. Soul consciousness is body consciousness. Our bodies are "physical" stuff full of conscious "is-ness" ever transforming. And it is within the power of individual choice to bring about more fulfilling life experiences for people living in relationship with each other. The physical nature of material subjects is qualified by their interactive properties. Their potentials of dynamic transformation are inseparable from their natural *quantification*; the "quantification" (objectivity) of subjects is but the physical configuration of their subjective potential for interactivity. Personally qualified consciousness seeks transformation that well serves the commonweal. We are the physical stuff of human "is-ness" and it is within the power of individual choosing to make choices that bring about more fulfilling life relationships. Hell on Earth is a consequence of human choosing. The devil is in the choices we make.

And, the *heart of the rose* is love. What is love? It is in the nature of personal rationality to develop one's own definition. For the Christian, Jesus Christ personifies the meaning of love. Love's best definition is not in words; rather it is in the example of right relationships personally chosen and lived. Love is harmony. The Christian mandate is to love God and one's neighbor as one's self. Our sense of the kind of relationships we want others to favor on us is the measure of how we should conduct ourselves in our relationships with others. The presence of God is best represented in the gratuitous providence of nature that gives us origin, that sustains us, and to which we return. The "holy other" of sustainable nature best embodies the grace of the divine; our reverence of God is best witnessed in our reverence for the Holy Other that provides for us, for all life.

At this point I might be forgiven if I return again to cosmological/philosophical jargon: we may say that love is the conscious rationality of cosmic purpose, and it is; we may say that love is the intentional energy of spiritual purpose that sustains the commonweal, and it is; but simply stated, if we always conform our appetites to the necessity of what is good for everybody, including the Holy Other, we live by love. Constantly striving to understand the consequences of personal choices, and to the best of our understanding and ability, to do things beneficial to self and others, and to avoid doing things harmful, exemplify what it means to love. Love is what reciprocally cultures beneficial relationships in context of the necessities of others, including the Holy Other, the global ecological continuum. It is the global continuum of network life that enables soul and body. Love is what makes peace possible between people individually and amongst nations globally.

Family is the place of love's social beginning. Love is the bond of *two-ness* reciprocally inspiring a will for *three-ness,* the option of mutual giving and living together as one. Love is the mutual option between a man and a woman (two or more persons) who choose to live in a relationship, the motive of marriage and all sociality. Simply stated, mutuality motivated in love for each other is *action that serves.* The essence of the "is-ness", of life, of cosmic purpose, is service. Quantum energetic consciousness serves purposeful transformation. Every day and every moment the purpose of love is real and personal. Godlikeness, true love, is expressed in living conformed to a mind of mutual concern in every relationship. Esoteric descriptions of love in dusty tomes of philosophy and theology are only rhetoric; the possession of books filled with profound rhetoric may massage egos, but the rhetoric of the printed word, no matter how grandiose, is meaningless until it converts to action in personal lives.

Wasted Earth is groaning for change. Human failure to live by love is destroying Earth, fragile life, including our own. Love is the transparent drop of nectar cupped at the heart of a rose's extravagant beauty—the scent of service. Mostly we go through life without savoring *the heart of the rose*. It is time to stop and to savor and to change.

The Sense of Trinity

Traditional belief in the Trinity Godhead has been fundamental in Christian Theology, as has naming God "Word". The *Procession of Word* in the Trinity Godhead is active; the beginning "word" is a verb. Word is the agency of communication, the Expressive Agency of and in creation. The purpose of Word is communication, the expansive means of self-expression; the purpose of communication is the joined agency of mutual consciousness in the expression of new consciousness, new word. Amplified and attenuated, consciousness evolves from the process of communication. In the evolved aggregate of associations, consciousness becomes qualified with new and differentiated potential. The cosmic continuum may be understood as the aggregating soul/substance of word procession.

First Community is expressed in the amplification and attenuation engaged in primal word communication. First Being is conceptualized in the communication of *two-ness*, wherein *three-ness* processes. The Primal Community of Three-ness expresses Godhead Trinity.

The purpose of communication is connection; communication's connection is joined consciousness. Consciousness is reified at the intersections of communicating entities. Consciousness is the fruit of dialog. In quantum-electric terms, plus/minus potential energizes, opens and enables new consciousness to amplify, attenuate and be attenuated.

The wholly new potentials existentiated by electrically joined potentials can be called the *purpose* of communication, the point of communication. It may accurately be said that all cosmic communication and consciousness are *quantum-electric*. The *purpose of quantum-electric communication/consciousness* is the expression of and reification of newness.

Non-consciousness is equivalent to nothingness for all substance is energetically (*spiritually,* in a raw sense) consciously qualified. Consciousness is associated with purpose, with *something-ness*; the absence of purpose equates with nothingness. Cosmic reality cannot advance on nothingness, which is to say, conversely, that it must advance on purpose. Quantum-electric attraction between the electrons and the nucleus, for example, is a purposeful agency that is appropriately called "consciousness". Incipient consciousness is "soul", becoming ever more subtle in ever more complex associations.

To call communication and consciousness "quantum-electric" merely describes the physical exchange of energy, the *physical* reality, the *quantum* thing; it says nothing about the diverse capacities for future creativity that is contained in the *qualifications* of quantum-electric soul/substance; the subject matter of "spiritual" qualifications has been left to the philosophy of *metaphysics* and to theology. The body/soul schism and the alienation of science and religion that have been caused by it have yet to be reconciled.

Patriarchal theology sees maleness as the First Principle in the Procession of Trinity, and on this presumptive basis patriarchy has evolved politically and theologically a male dominating hierarchy, which in its absolutist orthodoxy prevails in perhaps most religions, including Judaism, Christianity and Islam.

The presumptive primacy of the male in the human family, the primary social community of humankind, has given to evolved theology its paradigm of the *patriarchal human.* Theological consciousness today is beginning to challenge the presumptive *human paradigm* because of the all too apparent violence that stems from male self-arrogation over all other life, including its presumed divinely appointed right of dominion over women and children.

Insights of modern science into the cosmic ordering of atomic/molecular reality and the agency of quantum relativity in cosmic ordering, suggest that there may be a prior and more authentic paradigm, namely, that of natural *quantum relativity* by which atomic/molecular structuring occurs—the deep relativity of particulate energy which reflects a more fundamentally ordered relationship of male and female relationship.

"Maleness", the presumptive First Principle in the *paradigmatic human*, has, in light of social experience and modern science, lost its credibility as a governing model for conscionable human socializing. Orthodox patriarchy, in the harsh theology of literal application, foments wars, relegates women to the servitude of men and excludes women from equal participation in Church and State politics. By fiat of belief, women are denied equal rights under patriarchy's presumptive reading of divine law.

The *transformational paradigm*, the more benign model of evolutionary nature, is prior to the *human paradigm* for it goes back deeper in nature to the original quantum-electric relationships of the atomic nucleus/electron, wherein a better proportioned sense of the male/female relationship is discerned. The *elemental intension* that purposefully bonds (grounds) electrons (male component) to the nucleus (female component) better expresses male/female relationship and the spiritual analogizing of divinity Godhead. The communal ordering of Earth/human relationships in the continuum of atomic/molecular processing is driven by the trimorphic resonance of communication, consciousness and conscience.

The Evolution of Symbiosis
(By Monica R. Steffen—1982)

In past scientific experience it was thought that symbiosis had its place on the outer perimeter of evolution, displaying itself in such bizarre and exotic cases as lichens and cow rumens. It is now being discovered that in fact symbioses are widespread and central to the mainstream of evolution and are essential to existence. This discovery is of primary importance particularly in this day and age when polarity and contention are the guiding forces by which the highest evolved organism, Homo sapiens, lives in society. With the potential of human destruction looming dangerously on the horizon, I cannot think of a more crucial time to introduce this concept.

Symbiosis: What it means. The word "symbiosis" was first defined by the Botanist Heinrich Anton de Bary as being "the living together of differently named organisms." The more current definition is "the association, for significant portions of the life cycle, of individuals that are members of different species." But, according to Lynn Margulis who is a professor of biology at Boston University, "Neither the general discussions nor the definitions have been very enlightening, nor have they stimulated new research."

Perhaps looking back to the terms original roots without adding the scientific stipulations and limitations could provide a bit of insight into the essence of the word. Symbiosis comes from the Greek "sym" which means "with" and "bios" which means "life", so basically the word means "the state of living with." It is the interdependence of life. Symbiosis is a peaceful phenomenon as opposed to the two other related forms of interaction, pathogenesis in which one of the organisms is severely damaged, and parasitism in which one partner is gradually debilitated.

According to Margulis, "Symbiotic associations offer their members workable solutions to many of the basic problems of health and survival." Behavioral, metabolic and developmental factors combine to optimize the chances that natural selection act on the symbiotic association rather than on the symbionts as individuals.

Symbiosis: from the very beginning. There has been a revival of ideas with regards to the symbiotic origins of organelles in eucaryotic cells. This has arisen from present-day discoveries of the abundance of microbial symbioses coupled with the recognition that a plant or animal cell contains several remotely related, semi-dependent genetic systems such as the mitochondria and photosynthetic plastids.

Mitochondria, which are found in nearly all protist, fungal, plant and animal cells, are responsible for generating ATP by oxidizing the molecules obtained as food or produced through photosynthesis. They contain their own genes, which are similar to the DNA molecules found in viruses and bacteria, and they also contain their own complex internal membranes.

In cytochrome composition and in other physiological and biochemical respects, mitochondria resemble the free-living bacteria of the genera Paracoccus and Rhodopseudomonas much more so than they do the nucleocytoplasm in which they are found. However, mitochondria do not contain enough genetic material to produce all their own components and proteins and are dependent on the nucleocytoplasm and protein-synthesizing machinery of the cell. Such observations make it possible to believe that mitochondria were acquired as symbionts during the earliest appearance of the ancestral eucaryotic cell and persist in intimate association with their nucleoplasmic hosts.

Photosynthetic plastids provide yet another interesting example of microbial symbiosis. Not only do chloroplasts and other plastids have their own genetic mechanisms, but they also have a curious double membrane structure. The outer membrane is synthesized according to the instructions of the eucaryotic nucleus; the inner membrane is synthesized according to the instructions of the plastid's own genetic system.

Photosynthetic plastids bear a very marked resemblance in structure, macromolecular chemistry, and function in free-living, oxygen-producing photosynthetic coccoid bacteria. Of particular interest to evolutionary research today are the Prochleron because they have the features of predicted ancestral chloroplasts.

Symbiosis: A problem solver. As it was mentioned earlier, symbiotic associations provide workable solutions to many of the basic problems of survival. Providing adequate nourishment is one of the problem-solving advantages of symbiosis.

There is a species of green hydra, Hydra viridis or Chlorophydra viridissima, whose characteristics have been extensively studied. Their green coloration is due to their association with a species of green algae, Chlorella, that lives within the hydra's gastrodermal cells. As long as there is adequate sunlight, these hydras are able to sustain themselves with the food produced by their photosynthetic symbionts even if outside nourishment is not provided. The closely related white hydras, however, would starve without external provisions.

There are a variety of marine fish, which have developed a symbiotic interdependency with luminous bacteria. The fish provide sanctuary and food for these specialized bacteria, and the bacteria provide these fish with a steady, controllable light source for searching out food in dark waters and for signaling members of the same species.

Even some of our unwanted tenants, the termites, owe their existence to symbiosis. Dry wood and subterranean termites harbor an array of symbiotic bacteria and non-photosynthetic protists in their paunch. These symbionts break down the wood cellulose into forms that the termite can utilize. The termite ingests nourishment for the bacteria, and they, in return, convert these foodstuffs into nourishment for the termite.

Symbiosis also solves the problem of motility and large size for some organisms. Polymastigotes, for example, have a set number of cilia in relationship to the number of nuclei they have per cell. Selection pressure usually acts against an increased number of nuclei and therefore against good motility. In order to overcome this, motile bacteria are often acquired as external symbionts. These bacteria crowd the surface of the polymastigote and beat their bodies in synchrony with the cilia, thus good motility is restored.

Human beings too are about ten percent dry-weight symbionts. Bacteria are housed in the small and large intestines, on the skin, in the mouth, and in the genital tract. Many of these symbionts are responsible for synthesizing important vitamins and for chemically enhancing the digestive process.

As science explores the symbiotic phenomenon, it is becoming more and more evident that it is a common occurrence rather than a rare event, and according to Margulis, "...it is a product of an evolving inter-specific relationship."

Symbiosis: The human realization. The species Homo sapiens is the highest evolved organism on this planet to date, thanks to symbiosis and myriad other forces such as natural selection. Given the evolved human intellect, man has the potential to grasp the essence of his existence, to perfect himself, and to direct himself toward a purpose.

Within the last couple of decades, we have witnessed a knowledge explosion in all areas of science. This scientific period of enlightenment has given us much insight to our existence. It is now time to coalesce this knowledge and direct it toward a universal purpose. Man can no longer afford to continue his exploitive abuse of this knowledge without running the risk of destroying himself.

Man has yet to realize that in order for him to survive as a species, he must learn to live symbiotically with his fellow man, with other species of the earth, and with earth itself.

Symbiosis: A tool for learning. Science and religion provide vehicles by which man can realize his potential; unfortunately, they too have yet to be symbiotically united. In essence, science in its pure form and religion in its true sense are one and the same; their main pursuit is that of truth. The problem exists in the vocabularies of the two institutions. Their teachings are the same, but they are speaking two different languages.

Because of this miscommunication, misunderstandings arise, thus, polarizing the pursuers of the same truth. This polarity is found not only between the school of science and the school of religion, but it is also found within the individual schools. Within the religious institution there are many denominations with their many different beliefs, and those who hold a particular dogma would profess that theirs is the only truth. Within the scientific institution there are many schools of theory and thought, and those who pursue a particular theory in a defined manner without considering other schools of thought and other consequences run the risk of discovering only a half-truth.

It is not that these polarized institutions are all together wrong, for there is much truth in their teachings, but they are representing only a part of the truth and thus are only partly correct in their teachings.

Polarization between and within these institutions serves only to complicate the search for truth. Men of science and those of religion need to strive for a symbiosis of the two institutions; the language barrier must be bridged so those who seek to learn the truth may be taught a whole truth.

Symbiosis: Hope for human society. Our society, both national and global, is suffering from an acute case of the Old Testament's Cain and Abel Syndrome. The two brothers represent two contentious forces. Rather than trying to accommodate one another, they allowed their differences to create bitter resentment. The consequences of this animosity culminated in the demise of one and the long-term suffering of the other. There are many contentious forces acting on the world, which set individual against individual, culture against culture, and nation against nation. The consequences of these contentions and our failure to make accommodations are displayed daily in the many Cain and Abel scenarios that color the news media—and looming dangerously on the horizon are the consequences of a major contention between two nations that threatens the existence of all mankind and other life forms.

Perhaps this is the message of the New Testament's Book of Revelation. Taken literally, as many do take the Bible, this writing conjures up all sorts of unimaginable terrors. But if the mystical propositions are reduced to an applicable perspective, the choices it offers and the consequences of each choice may be brought to new light.

Apocalypse (from the Greek "apokalypsis" which means "uncover") may be said to represent the uncovering of the true purpose of our existence, the symbiosis of all nations such that a unified, interdependent world society is created.

Armageddon, the antithesis of apocalypse, may be said to represent the consequences of pursuing a contentious purpose, the maintaining of individual nations such as to perpetuate a divided world.

Unfortunately, we are currently living in a very divided world, a contentious world with a nuclear holocaust impending, and man is still applying his evolved intellect and acquired knowledge to self-serving purposes. He has yet to learn to be accommodating, to become symbiotic. He is still parasitic and sometimes even pathogenic in his relationships to other men, to other forms of life, and to the earth. But even so, there is still hope for mankind; according to Margulis, "...parasitic relationships may become benign and even obligate..." and thus these relationships would become symbiotic.

The laws of love that are taught by many of the world's religions, Christianity to name just one, are the laws of symbiosis. When man learns to become accommodating, when he learns to live interdependently, then would true religion and symbiosis be experienced.

We have seen how symbiotic relationships between organisms have been used to solve problems of survival. Human symbiosis may well be the key to solve many of the problems that face mankind. But in order to take on such grandiose tasks as achieving world peace and feeding the millions of hungry people in this world, we must first learn to be accommodating in our daily existence. We cannot become a symbiotic world society unless we learn to become symbiotic individuals.

It is time for us to realize that we live on a physically finite planet. The material things we borrow from this earth in order to survive are precious and must be shared. Yes, symbiosis is the way.

Footnotes:
1. 4-18 Lynn Margulis, "Symbiosis and the Evolution of the Cell,'1982, Yearbook of Science and the Future, Encyclopaedia Britannica, Inc., 1981, pp. 104-121
2. M. P. Starr, "Bdellovibrio as Symbiont," Symp. Soc. Exp. Biol., 29, 1975, 93-124
3. Lynn Margulis, "The Genetic and Evolutionary Consequences of Symbiosis," Exp. Parasit. Rev., 39, 1976, 277-349
4. Special reference: my father who has given me colorful insights into religion, science and life.

The Evolution of Consciousness

Consciousness is a quality (faculty, grace) of awareness, a *spiritual* quality that evolves. Aspects of consciousness are the codependent and co-evolutionary qualities (faculties) of Reason, Faith and Purpose—the main divisions of this book. These are natural correlatives and products of the trimorphic processes of communication, consciousness and conscience.

Self-awareness, awareness in general, makes up a unified consciousness born out of the codependent graces of reason, faith and purpose, and their agencies of communication, consciousness and conscience. The grace of consciousness is a quality of soul that characterizes the individual personality. Mind/body personality, embodied consciousness, is a unity of soul/substance, a continuity derived from Earth Personality, from Cosmic Personality—a continuity of starlight and stardust in continuous process of amplification and attenuation. Human consciousness resides in the mind/body venue of the individual and of the collective. It is an integrated rationality made up of remembered experiences, associations and extrapolations. In the venue of consciousness, the experience of relational value is a moral unity that knows no division in fact, but which is both "secular" and "religious".

Self-aware purpose, rightly or wrongly motivated, drives destinies. Good and evil are outcomes of driven destinies. The challenge to every person is to discern motivation in choices so as to prefer options that have good outcomes rather than bad. The dilemma is that virtually every action has mixed outcomes, which become clear only with time; thus, the need of an awareness of the ambivalence of choice and of the need to monitor outcomes and to adjust choices so as to minimize evil and maximize good.

This writing is sectioned in the natural sequence of the co-evolved faculties of reason, faith and purpose. This arrangement flows naturally from the threefold division of Primary Scripture, namely, communication, consciousness and conscience. This division is "rational", that is, a convenience of understanding (*ens rationis*); it is not "real", for natural complexity implicates all the above into the singular vital fabric of internetted unity and continuity.

Stated briefly, this distinguishes *transformational* consciousness (worldview) from *static-centrist* consciousness. The transformational worldview (consciousness) is *spiritually* forward-focused on motive (soul, cosmic spirituality), while the static-centrist worldview is backward-focused on past things, scripted events, and objects. Transformational consciousness recognizes the ambivalent good and bad in everything, while static-centric consciousness characterizes objects, things, as good and evil, when in fact they are in themselves indifferent. Motive alone is good or evil or undecided; motive "qualifies", that is, determines relationships with, and uses of, objects. Herein lies good and evil, the outcome of human choice.

The cultured division of the realms of the "religious" and the "secular" has introduced a real schism in the unitary venue of consciousness. Social redirection must be premised on the integrated sense of common consciousness, for societal dissolution is certainly in part at least a by-product of cultured alienation. The reconciliation of religious and secular consciousness, religion and science, seems on its face to be urgent and self-evident. The rift between science and religion is so deeply cultured that their reconciliation will not come easily for vested interests run deep. So it is appropriate that what is intended and what is not intended by efforts to bring about their reconciliation be clarified.

The endeavor to reconcile science and religion, the religious and the secular, intends neither the secularization of religion nor the institutionalization of state-sanctioned religion. Nor is there any intention of demeaning faith practices or of discrediting religious ritual. Simply stated, the reconciliation of science and religion intends the ongoing integration of gathering knowledge into the moral framework of human consciousness. The reconciliation seeks to provide a framework of harmony where the myriad facets of evolved consciousness can be assembled into a rational collage of resonant significance in individual and collective venues.

Mary Settegast's book, "Mona Lisa's Moustache", ("The Ever-Present Origin", Athens, Ohio University Press, 1985) deals with the evolution of consciousness. She acknowledges the influence (pg 124) of Jean Gebser and his insights on evolutionary consciousness. "Gebser saw human consciousness as an evolution, an unfolding through five historical stages that he termed archaic, magical, mythical, mental-rational, and (now emerging) integral structures of consciousness...The incoming integral consciousness allows, as the name implies, the free expression of all the other structures without being captured by any one of them...There are people practicing shamanism and other forms of magic, while some of the world's oldest organized religions worship mythical concepts of gods and goddesses, and theologians argue in mental-rational terms about ethics and the nature of God. But a new coalition of these structures of consciousness is also observable in our time". (Pp: 108-109).

The emergent consciousness is from the wholesale deconstruction and dissolution of structures that have occurred especially in the Western cultures over the last several hundred years. One likely public reaction to the pessimism attending dissolutions is to fall back on fideistic crutches and "to retreat into fundamentalism". An alternative to shifting from one dead end to another, recommended by Gebser, is to pursue the "open quality of the world through our own openness." (Ibid Pg: 133).

The astronomer Johannes Kepler and the philosopher of history GWF Hegel after him subscribed to Joachim à Fiore's division of history into three eras corresponding to the archetypes of the Persons of the Holy Trinity; the era of the Father (paternalism and patriarchy), the era of the Son, second coming, and the third, the era of the Holy Spirit, the socializing Spirit of love and the conviction of conscience. In this division, resonance with Gebser's five consciousness stages of history can be found. The "archaic, magical and mystical" correspond with paternally, patriarchically conferred faith and fideism; the "mental-rational" stage corresponds with the Second Person consciousness of the Son, hope; and the stage of "integral human consciousness" corresponds to the conscional awakening of love under the discerning consciousness in the age of the Holy Spirit. The 2000 years since the time of Christ can be considered the historical stage of the Son; Old Testament history prior to his coming can be considered the stage of the Father; and this era, the *dawning of the*

Age of Aquarius, can be considered the stage of "integral consciousness."

Femininity
Cosmic Rationality's Ground State

As the atomic nucleus monitors and grounds the energetic attraction between the nucleus and electrons, so woman is the central consciousness and grounding of the nuclear family; the elemental *trinity* of these essential structures, physically, psychically and vitally, is embedded in all societal groupings. Conscious fidelity by all Earthlife groupings (and by individuals composing them) to the essential fact grounding of their vitality in atomic/molecular "femaleness" is prerequisite for individual/social vitality to exist and flourish. Self-reflective consciousness enables reflective intention and purposeful action. Femininity, consciously grounds free electrical energy (electrons—masculinity) in purposeful action. Femininity mirrors divinity directly in purposeful creativity, for it is the evolved rationality of cosmic vitality and the inherent centeredness of family, Church and civil society. A perhaps workable definition of femininity is *the evolved ground of conscious, cosmic soul/substance*.

If a fundamental cause is to be identified for the wasteful disasters being wreaked on Earth and her vital resources, a prime candidate is the violent effrontery of males who have from time immemorial arrogated unto themselves—to the exclusion of women—the ground state position in the organization of societal order. Hope for humankind, indeed, for all life, lies in a turnaround of societal consciousness that restores female centeredness to its universal role in the cosmic working of atomic/molecular soul/structure.

Personal call to conscience is inspired by family experience. When family voice speaks authentically in the upbringing of a child, the child can most generally be trusted to respond authentically. When family voice lacks authenticity and speaks from conflicted grounding, the child will grow up conflicted. Modern societies are composed of members who have grown up in a world of conflicted voices. These members have been deprived the experience of social/familial harmony and are uprooted from a consciousness of their own authenticity. This psychic disorder, that is, the mental confusion of being uprooted from one's inherited authenticity by conflicted voices, is called "schizophrenia". So imprinted in societal patterning is male arrogance against the cosmic authenticity of female grounded-ness that the psychic disorder resulting from it is manifest in physiologic disease—to the point that it is now difficult to distinguish cause from effect. Indeed, certain genes may now be transmitted which physiologically predispose certain individuals to the manifestations of schizophrenia.

In authentic family living, the call of conscience, if not instinctive, certainly appears early in individuals as a sense of personal obligation to serve one another in the family setting. This *instinctive/intuitive* rationality is a "rationality of service", the underlying rationality of the universal call to *priesthood*. It is as essential to female rationality as it is to male, if not more so, for nurture is experientially and preeminently associated with femaleness. Priesthood, the call to service is quintessentially the conscionable intuition of cosmic vitality—femininity.

Life's expectation received from the lineage of all life that preceded us is a natural inheritance. It's the same for every person and is really quite simple, as St. Paul realized: "to proclaim the truth openly and to commend ourselves to every man's conscience before God." (2 Cor: 4, 2). As St. Paul says, "this treasure we possess in earthen vessels". The treasure's incomparable worth finds its measure in Jesus' example and teaching. "For God who said, 'Let light shine out of darkness,' has shone in our hearts, that we in turn might make known the glory of God shining on the face of Christ." (Id: 6,7). That treasure is the personal consciousness of divinity; personal consciousness finds its worth in the lived lessons of exemplification and teaching. If we do not exemplify and communicate the message of Jesus, Christian altruism, we miss the very service-mission that characterizes the essential grounded-ness of our own being. Paul's Christian insight of universal obligation to conscience is at the same time the rationality of cosmic consciousness. Exemplification and teaching live on even after our "earthen vessel" has yielded its light to death and its fragile tissue has returned to Earth's dust whence it came. Light, insight, consciousness, conscience, cosmic/Christian rationality/spirituality, is the message, and we "earthen vessels" are its messengers. In fact, like Jesus, like Paul, we are the messenger and the message. Female and male, we are, each and together, messenger and message, priest and "eucharistic" victim. Like wheat, we are literally gifts of light; we are bread and spirit—Eucharist—to one another. Maturity is a dying, the shedding of our own light for the benefit of others. "Unless the seed dies, it remains itself alone." Light is our origin. Light is our destiny. The sharing of light is our mission.

My personal recollection of family experience reinforces my awareness of the natural rationality of service that was nurtured in the family of my upbringing. I was born (1933) the second youngest in a family of ten children, so I experienced the altruistic support of a large and loving family. I was a secure and happy child. As far back as I can recall I wanted to be a Catholic priest. Although we lived on a farm, we were fortunate in that a public country grade school, Barclay No. 3, was located at a country crossroad across from a Catholic country Church, St. Francis Parish, Route 1, Dunkerton, Iowa. I had an older brother Arnold, who expressed also his intention to become a priest. As it has turned out, my two sisters both became Dubuque Franciscan nuns, and Arnold became a missionary priest (with the Society of the Divine Word, SVD), still in New Guinea since 1958.

In 1946 I matriculated at the SVD Preparatory Seminary, Epworth, Iowa, and I continued with the SVD until September 1957. At this time, having completed High School, Novitiate, College (Juniorate), Philosophy and First Year Theology, I came to the decision to discontinue formal studies toward becoming an institutional priest. In a sweet turn of irony, I went on to marry and became the privileged father of six daughters and grandfather to seven grandchildren. The reason behind my decision not to become an institutional priest, as I gradually have come to understand it, was my conflicted consciousness over institutional Catholicism's inability to reconcile universal and institutional priesthood with masculinity and femininity. The institution advances a theology that alienates women from it, not only from priestly service in Church, but also from its rationalized concept of God. This seems to me a very elemental misdirection

The intuitive sense of harmonic family is to predispose young people to vocations of altruism. If their experience in responding to their intuition discovers institutional conflict with natural harmony, the young are likely to opt fidelity to natural harmony over the conflicted voices of institutions. This may explain, in part, today's crises in vocations to the priesthood and ministry; namely, that institutions fail to ground themselves in the rationality of natural authenticity and fail ultimately in sustaining their appeal to youthful idealism.

The centrist, male mindset of Roman Catholicism is so engrained, theologically and structurally, that is unlikely either for its theology or its male hierarchy to change unless and until women *en masse* refuse to accept the *status quo*, and, society as a body claims its natural, God-given authenticity, theological and structural. After all, it is by reason of divine intention, scripted in Christian and cosmic rationality, in femininity, that Sacrament is experienced universally in the intention/intension of cosmic energy and matter—in conscionable soul/substance.

Women Renewing Church

For many Catholics today, perspective on Church and women's role (or non-role in it) has been shaped by events during and after Vatican II. As the father of six daughters, all born from 1960 to 1969, I have been formed spiritually not just by Vatican II but also by the experience of family theology. Conscientious parents seek to alert their children to risks that threaten their well-being, and this may include—which was unthinkable before Vatican II—real and imminent risks that are associated with the cultured traditions and regulations of institutional Catholicism. What these are may be discerned in the contrast of the spiritualities of the First Vatican Council and the Second Vatican Council.

My wife and I were reared in the hierarchical, male micro-managed theology of Vatican I, and we were ready for the freshness of Vatican II, which put greater responsibility on the individual in the formation of personal conscience. My wife was college educated under the Maryknoll Sisters in Manila, Philippines, while I spent eleven years (1946-1957) in preparation for the priesthood with the Society of the Divine Word (SVD) in the United States. We met as graduate students in 1958 at Notre Dame, Indiana, at a meeting of the Newman School of Catholic Thought. She completed her master's degree in teaching at the University of Northern Iowa, Cedar Falls, and I completed mine in Botany (Plant Physiology) at Iowa State University, Ames. We were married in June of 1959 at the St. Thomas Aquinas Student Center in Ames.

Informed, personal conscience is the supreme authority compelling individual living. Conscience is always in the specific, for personal choice is a matter required in the specific. The chore of informing personal conscience is an unending work because the circumstances of daily choices, the spectra of their outcomes and consequences, are interdependently complex and mostly not anticipated by institutional prescription. For example, the term "right to life" has personal as well as social implications. The morality of *right to life* needs to be evaluated with respect to the way personal and social implications relate to each other. Significance respecting the individual person pertains to the moral right of the conceptus to life; significance in the social perspective must also include considerations of the mortal burden humankind has become globally in overwhelming network life, and network life's loss of capacity to sustain future life. The mortal degradation of network life and its environment, upon which the future life of humankind depends, is an abortion as surely as is the mortal termination of an individual pregnancy. If this is true, then population control becomes a proportionally compelling moral issue of personal conscience. "Birth control" is not an issue that can be disposed of by the pronouncement of some autocratic, male institution. On its face, it seems obvious that the celibate, male priesthood of institutional Catholicism is hardly an unbiased authority in such a matter, for priests, by their own choosing, have removed themselves from the need of choosing personally whether or not to have a child. The authority of the all male priesthood is especially in question because it excludes women from all decision roles of social morality, and thereby excludes from its decisions the very ones who bear the children. The imbalance of this, on its face, seems to demand correction.

New women in new Church know who they are and who men are. Women are not merely "wom(b)men", they are unique and conscionable individuals in their own right, who owe obeisance to no authority less than God and their own consciences, whose rationality derives from the female/male covenantal consciousness in paradigmatic nature. Marriage is defined not by contract but by societal covenant in which male/female mutuality in interpersonal relationships obliges man and woman to self-commit to communal conscience and well-being. If communal conscience and well-being are unilaterally compromised, so are individuals, families and society itself.

It is the new woman's sense that the place of religion is everywhere in conscionably lived, interpersonal relationships. More than the orthodoxy of instituted belief and practice, religion is nothing if not a self-aware inner disposition that orders the word and work of personal and communal self-perfecting. In their communal aspect, women and Church are essential codependents, that is, they witness publicly the "re-membering" of the sacred—in the literal sense of the genesis of new life and renewed living, and in the authenticating sense of celebrating in common sacrament that which edifies personal and communal whole-making. As nurturing mother, woman and Church *re-member* and teach "to remember, to celebrate and to believe".

New woman's mind is not imprisoned in the regressive and pessimistic fixations of an ancient, patriarchal theology, rather it is progressively and optimistically oriented toward present and future well-being. Her theology is not psychologically confined to instinctive hard wiring, rather, it is open to the dialogical intuitions of the cortical brain. Her consciousness of "service priesthood" is energized in her open intuitions, not nullified by male, exclusivist proscriptions. Her love is unconditional and non-sexist.

132

Male-instituted Catholicism seems, after the mind of Peter, to have politicized itself in the corporate culture of prevailing mercantilism. Corporate mercantilism thrives by locating itself apart from other ideologies. Corporate business, for example, knows that it needs to have opposites against which it can argue its merits. Polarities are good for business. If corporate business doesn't have a "devil" against which to compete, it finds ways to create one. In the natural course of events plenty devils already exist and new ones don't need to be invented. The culture of polarities tends to alienate people from the middle ground and to concentrate them at the warring poles. Societal harmony requires people to accommodate themselves to one another and to mitigate fractious polarities. Institutional Catholicism has evolved exclusive rituals and dogmata, which are calculated to distinguish it from those characterizing other Christian Churches. As Jesus in his time chased the temple merchants out of the temple, so he would today. His gospel and work are not marketable commodities, rather, they constitute a philosophy of life and a commitment of faith in the Naturalis Sacramentum Ordinis—the sacrament of natural (cosmic) order.

New Church will structure itself systemically after paradigmatic nature, i.e., it will be morally inclusive of the whole network of vital interdependency. Human vitality and religious consciousness are totally dependent upon network vitality and upon male/female mutuality. The adversarial model of institutional religion has depended for its mercantile success on the culture of polarities rather than on the reconciliation of polarities. The fact of cosmic, psychological polarity is a paradigmatic, natural fact. But the fact of natural mechanisms (symbioses), which account for life's success on Earth, is also a paradigmatic, natural fact. Even before self-aware consciousness existed in the cosmos, nature found ways to mitigate fractious polarities. Human beings originate from the continuity order of the universe, and as agents within that order they are morally and practically responsible for sustaining it, for resolving psychological polarities.

New Church will reflect more the mind of New Woman. It will not be mercantile, exclusivist and narcissistic. It will be inclusive and open. It will identify priesthood as dedicated self-commitment to universal service. It will not be politicized in an exclusivist bureaucracy of control. It will liberate and authenticate individuality. Its paradigm will be symbiosis; it will seek to mitigate polarities; it will avoid hyping them. It will affirm differences and explore them for mutually serving benefits. While it will engage in a universal logic, it will not entrap minds in monopolized ideologies. It will mirror the mind of Christ, the Alpha/Omega of evolutionary consciousness.

Life's complexities present urgencies, which require choices. The consequences of indecision in the face of morally compelling urgency can be grave. A contemporary and compelling urgency is to choose which worldview best motivates personal and public consciousness, the biblically acculturated absolutist-male static-centrist worldview or the contemporary transformational worldview of quantum science. We are taught that God more than frowns at indecision, God "vomits out" lukewarm-ness. More reprehensible than indecision is the intentional choice to remain uninformed in difficult matters, even though they are morally of critical importance. By intentional negligence the middle is pitted against the opposites, and contentious irresolution results. Intentional ambiguity seems to be the calculated choice of Church professionals who adhere theologically to the absolutisms of the ancient static-centrist worldview while professing openness to the cosmology of quantum science. Quantum rationality does not accommodate static-centrist theology. The stultifying consequence of indecision is inaction and apparent ignorance. The product of calculated indecision and inaction is to be neither "hot nor cold", but to be "lukewarm", which theologically is to be morally at fault. The patriarchal theology of static-centrism lacks moral grounding in transformational consciousness and links directly to the consumerism of modern corporatism. New women in new Church, while cherishing and affirming authentic masculinity, know better and speak out prophetically against the inauthentic system of patriarchy.

Opinion of a Twelve-Year-Old
(Monica Steffen—circa 1973)

"The Roman Catholic Church is very slow to change. Many parishes still follow the set concepts of Fifteenth Century religious standards. Although Masses have been translated from Latin to English, they are still performed in basically the same manner. While the world outside is learning to see men and women as equals, the Catholic Church has refused to allow women to become priests. This must change! The Roman Catholic Church should allow women to be ordained into the priesthood.

"God's call to "man" is to serve Him by helping our fellow "men." (I use these in the broadest sense of the terms.) There are many different capacities in which one may choose to serve Him, and one should not be limited just because of one's gender."

[*Monica was broadly talented, of strong moral conviction, and totally lovable. She would not be deterred from getting a well-rounded education even though the politics of college curricula fought her all the way.*]

Sense and Insight

With respect to what was going on in Roman Catholicism during the Second Christian Millennium, it seems accurate to say that the "Modern Period" (Modernism) roughly coincides with the spread of mercantilism and the money/power corruption within the Church (the Holy Roman Empire-Europe); especially, for example, with the ascendancy of the Florentine de Medici merchants and bankers to power and authority within the Roman Catholic Church (and Europe) over a period of approximately 300 years. Mercantile corruption and the global outreach of colonialism, driven by motives of financial exploitation, fed the movement of European feudalism and extended the political domination of princes and prelates. This corruption led to gross public abuses, which widely fomented public protests, upheavals and demands for Church reform.

The rise of Protestantism was a tide that could not be contained because the best religious instincts were being massively trashed. Political and social violence became the law of the day. Inquisitions and crusades were initiated and were especially violent in their zeal for they were rationalized and executed "in God's name".

The domains of science, philosophy and theology were venues in which wars had connection. After Copernicus, science contributed seriously to the psychological disabling of the long hallowed mindset of cosmological/theological/political centrism, whether it pertained to arrangements of heavenly bodies with each other or whether it pertained to the theological/political domination of people by monarchical Church/State.

Philosophically, Modernism is characterized in upheavals of thought and the many-faceted dissolutions that came from the upheavals. Niccolo Machiavelli (1469-1527), credited with being the first philosopher of history, gave to Roman princes and prelates the political philosophy they needed in order to carry out the feudal domination of the masses. He was a contemporary of the Dominican Monk Girolamo Savonarola (1452-1498), who was a bitter enemy of the Medici rule and Church corruption, and who confronted the corruption of the hierarchy and was martyred for his efforts. Later German philosophers, many good Lutherans, the likes of Spinoza, Kant, Leibnitz, Goethe, Schelling, Schlegel, Schleiermacher, Feuerbach, Nietzche, Hegel, Freud and others, gave impetus to insights that further eroded old and settled beliefs and practices, and contributed to the casting of a heavy blanket of pessimism over the established and once comfortable European (Roman) world order, Church and State. Protestant scientists, philosophers and theologians (former Catholic priests and monks) contributed to the chaotic dissolution of the old world order that was once based on the philosophical and political rigors of static Earth, centrism and absolutism. The dissolution seems to be accomplished with the philosophical pronouncement that "God is dead", that is, the old centrist, politically absolute god of the Holy Roman Empire was no longer acceptable.

The "Postmodern Era" begins, burdened with the task of making sense of the dissolution it inherited from Modernism, and with the task of creating a new world order out of the ruble of imploded Romanism. The task is not only unfinished, it has hardly begun, for the old centrist philosophies/theologies still dominate in the political theories of Church and State. Presumptive cosmological centrism has not yet been dislodged from public thinking, nor has the presumptive theological reasoning that evolved from it yet been seriously challenged, let alone replaced in the consciousness of the powers that be.

It is here respectively insisted, that just as the centrist worldview of times past served the knowledge of the era and its institutions, so now, a worldview adequate to the knowledge of this time is needed to restore credibility and effectiveness to modern public and religious institutions. It is further suggested that the cosmological consciousness of the quantum-electric universe (the transformational worldview) can give global societies a politically more harmonious rationale (democratic) that is inclusive and egalitarian, and to religious consciousness a more open vision of God whose "center is everywhere and whose circumference is *nowhere*".

People of art, rhetoricians and philosophers, might play with the word "nowhere" and open its meanings to more positive interpretations and split the composite word a little differently, for example, "now here". In the light of the principle of *isotropy* we are informed that the kingdom of heaven is "now" and "here", for no matter where in the cosmos we might be, the kingdom of heaven is *then* and *there*, the same as *now* and *here*. (This insight is first attributable to the Dominican Monk Giordano Bruno; see the essay herein, "Pursuing Truth".) It is all a matter of perspective, whether the focus is on "the tree or on the forest". Tree-focused consciousness needs to become more forest-focused.

The function of the senses has everything to do with consciousness and insight (perspective and perception), in two meanings: 1.) physically, in the sense that cells and organisms "know" how to function harmoniously for the body-system's welfare, and 2.) psychically in the sense of intentionally informed and informing consciousness. The process of informing consciousness is a process of dialog that is at the same time physical and psychical. The physical "informing" aspect activates structures and structuring. Sense and insight are deeply interior tracks of perception. "Sense" pertains physically to the communications of the five senses informing commonsense, the reflex, mind/body response of patterned knowledge/belief engaged in the ongoing processing of experience and new information—that which is "immanent". "Insight" pertains to purposeful consciousness, consciousness of connections, the making of connections of faith and information, and corrections when knowledge and belief are discovered to be misdirected and/or lacking—pertaining to the "transcendent". Sense and insight pertain to physical and psychical correlations and to the securing of faith's certainty.

The path of neural tracks that carry the messaging of faith (emotion) and insight (reason) leads into the conscious field of "depth perception". Consciousness, as a complexity of vision, is binocular, emotional/rational. Perhaps the main distinction between vision and consciousness is depth of perception. Depth perception can be distinguished between the "objectified" depth and the "primordial". [Laura Sewall, "Sight and Sensibility", 1999, pg. 167, Jeremy Tarcher/Putnam, NY] The stored cumulus of experienced sensibility is wisdom's depth consciousness. We individually receive this stored consciousness potential with our parentally received genes. However, as with every genetic proclivity, consciousness is not automatically self-expressive; its expression happens in stages and depends on hierarchical patterns of development. If windows of opportunity at developmental stages are missed, then consciousness may be permanently and irreparably crippled. Consciousness needs to be cultivated pro-actively for negative culture agitates negative proclivities and positive culture nurtures positive proclivities.

Gazing across an open field, or from horizon to horizon across the open sea or desert, binocular vision helps focus consciousness to the center of one's being. Turning full circle, the visual mantra defines the importance of being by putting one in the center of one's immediate world, no matter where one stands. The depth and density of eternity is everywhere the place of conscious belonging. When we fix our vision on the deepest point of the horizon and walk parallel with distant objects, objects in the foreground "move" in and out of our vision while our gaze remains fixed on the distant background. If we circle a background object and keep it in our vision, then it becomes the visually experienced center of our world. *Objectified* vision attends to the flow of immediate objects while *primordial* vision sees passing objects but keeps the background focused. Primordial vision (the *forest*-focus) reinforces self worth by allowing self to recognize its centrality in relation to its immediate surroundings. Consciousness of the two perceptions is reconciled when the primordial is recapitulated in the objectified (*tree*-focus). Consciousness "rises" in personal experience (self-awareness) when vision of the near relates to the distant.

Depth-consciousness, even as physical vision, travels on two points of reference; its two points of vision are faith and reason. The vision of one "eye" is processed reflexly with reference to information provided by the other "eye". Depth perception is a function of many points of reference. Left eye information is processed in the right-side brain and right eye information is processed in the left-side brain, even as faith consciousness is considered more a function of one brain side, and reason, of the other. In visualizing space, artists exercise right-eye/left-eye perspective and valuate negative space in context of object occupied space. Artistic vision grasps space perspective more as a correlated, intuitive insight of object relationship rather than as a logical, rational one. Authentic reason and authentic faith coalesce in informed consciousness, which moves in the direction of the common purpose self-authentication. The two tracks of faith and reason converge on a common destiny.

At a given point, whether one looks forward or backward, faith and reason converge in primordial consciousness. From the primordial perspective, moments (time) and objects (place) between Alpha and Omega, between the back horizon and the front horizon, between life's beginning and ending, are isotropic, that is they are in many ways very much the same because they are in fact instances and objects of common, transformational soul/substance. What is important to perspective is that faith and reason keep parallel in their progress on time tracks. If one gets ahead of the other, their purposes may work in conflict and *distract* (derail) the train of consciousness from its purposeful progress. Integrated consciousness (reason-informed faith and faith-informed reason) essentially depends on the mutual functioning of both tracks. Faith is the grounding of consciousness informed in past experience; reason is the adjustment process that keeps faith on track in the face of object-thickets.

To press the analogy further, faith may be understood as a patterned matrix against which incoming information is situated according to patterns of experience. The stable text of established pattern is quintessentially female, while, fluid, contextual information is more characteristically ephemeral, male. Males more obviously deal with the objectified, while female stability is more primordially connected; males are more flippant, more distracted, more fleeting in their engagements, whereas females handle fleeting, short-lived events with a gravity that is more long term in perspective. Lasting relationships are important to females, less so to males. The eye of faith is trained in the relevance of things that have primordial depth, whereas, the eye of reason tends to be more object-held, distracted by what is close at hand.

We might say that faith functions in parallax; its vision takes in the whole distant scene even as it is aware of the changing perspectives of transient objects close at hand, but is not distracted by them; like a passenger looking out the window of a speeding train, faith's panoramic view is not distracted by the rush of telegraph poles just alongside the train. We all live in the primordial deep, exposed to a rushing tide of objects.

Perception, balance and authenticity are each a consciousness of a kind. For purposes of our consideration it is useful to understand perception from two perspectives. The first is a capacity to valuate relationships, recognizing the difference between short-term values and long-term. The second is a reflective dimension, the capacity of intuitional consciousness (patterned insights) to correlate valuated information against faith-patterns that are lasting.

From the perspective of personal experience and self-knowledge, we know, that different as these functions are, they are naturally unified in consciousness, they are interactive, and they are both necessary for self-development and fulfillment. The valuation of relationships, and actions proportional to valuated relationships, is true wisdom, the essence of primordial life's purpose. Consciousness is female/male characteristic—in the image and likeness of God—motivated in love. Maleness matters in femaleness.

Intuition and Speculation

If intuition is in fact an accessible well of deep-past information, encoded in personal/social consciousness, then there must be some way of tapping into it and using it to good purpose. Intuition is understood here as a faculty of common consciousness—commonsense—a personal and social faculty. The substantive *nature of things* is, as presented here, cosmic intension—"natural" sacrament, and the substantive *nurture of things* is cosmic intention—"nurtural" sacrament. The rationality of intention (nurture) has the power of giving direction to intension (nature). Perhaps more than ever, humankind needs to plumb the wells of intensional/intentional powers (nature-faithful conscience) in order to be more authentic and less wasteful and self-destructive.

The process by which humans might do this is the process of speculation, that is looking into and drawing upon all the experience and collective knowledge of humanity, identifying destructive and constructive understandings of actions, avoiding destructive actions and doing constructive actions as suggested by valuated understandings.

Faith and reason are called to steer consciousness with purpose as it negotiates the divergent journeying of destiny and desire. Integral human consciousness is a steamship named *desire* on the sea of infinite consciousness, a secured vessel in open waters. Intuition is a current of purpose tracking through chaotic seas. Active rationality is a train straining to carry passengers and baggage on safe tracks through uncharted vistas of destiny and desire. Steamship and train both return to harbor and station and swell the tide for future comers. Humankind is a complex of steamships and trains crisscrossing seas and plains in pursuits of desires and destinies inflating imaginations.

From the evolutionary principle of *essential continuity*, it might logically be derived from the *a posteriori* perspective: *that purpose in cosmic activity is patterned along pre-directed pathways toward some pattern-driven destiny.* The backward tracking of consciousness down the pathways of causally prior agencies concludes to a rationality of purpose along the many diverging and converging paths traveled in the past by intensionally/intentionally driven consciousness. The principle of the evolutionary connection between cause and effect substantiates the conscious logic of purposeful transformation.

Chardin coined terms, which suggest the genetic transformation of progressive purpose, for example, *cosmogenesis, anthropogenesis, Christogenesis and Noogenesis.* One might see in cosmogenesis and anthropogenesis the transformational linking of the cosmic process to Earth's vital process. The conscious progress of transformation is from anthropogenesis to Christogenesis, from self-centrism to altruism. Christogenetic insight is understandable as the consciousness of "second person coming", and Noogenesis may be understood as the fusion of integrated consciousness—fullness in the coming of the Spirit of love and conscience. In the present complex of humanity, havoc and violence will dominate except *intentional love* becomes a common bonding of purpose, a mental consciousness that motivates the collective exercise of integral human consciousness.

Divergent consciousness is disruptive of social harmony except it conforms to some conscious framework of universal integrity. The whole of cosmic consciousness, *the sea of infinite substance*, is a singularly integrating ocean of conscious vitality. The fragmentation of consciousness into divergent, irreconciled fields puts consciousness in internal conflict, vitiates holistic purpose and inflicts violence and degradation. Misdirected "professionalism" by the topical fragmentation of consciousness contributes to the frustration of integral human consciousness.

Imagine diverse knowledge as many schools of fish in the sea; then imagine, for example, if all the diverse life forms populating the world's oceans were isolated from each other, and that each kind was made to live isolated with its own kind. Havoc would invade all forms of life. In the attempt of each kind to survive isolated with its own kind, the strong would become predatory and impose themselves on the weak. A hierarchy of domination would quickly be established wherein the strong would dominate and cultivate the weak to serve the well being of the strong. Slavery! The internecine violence of struggles to survive would be awful. Isn't there something familiar about this scenario? Culturally, it's what humankind tends to do. Isn't that something like what happens when the manifold and interactive fields of consciousness, e.g., religion and science, are professionally appropriated and isolated? By now it should be obvious that common "religious" consciousness cannot be isolated from "secular", and vice versa, except with great violence to both; thus, the call for their reconciliation, their coincident elaboration.

The success of humankind, like the success of evolution, is the edification upon which the success of life is structured. The crisis of global life is the self-alienation of humankind from the universal community, which is mortally violates the networks of co-evolved life.

The creative expansiveness of life on Earth seems to have reached an impasse due to the aggressive and dominating self-assertiveness of humankind. Humankind has become nothing less than a catastrophic malignancy, a cancer wasting other life and threatening its very own. The massive destruction of species now in progress is nothing less than an irreversible abortion perpetrated by human arrogance and crass disregard for essential codependency.

Is there a way out? Cosmic transformation becomes transparent to human consciousness when it is liberated from myopic time-concerns and when patterns of archetypical forms communicate and resonate with integral consciousness in mutually inhabited spaces. Only then can variant, ephemeral perspectives harmonize and advance on structures of consciousness that are holistic.

Any fixation of consciousness in disconnection of the immediate present is but a time-stopped frame of momentary reality. Because we are wholly integrated into the fact of the evolutionary process of cosmic consciousness, we carry within ourselves the necessary bits and pieces of information that complete the mosaic of consciousness. We are personally but a facet of the great mystery of consciousness, and because of the small bit that we are, our self-understanding is incapable of grasping all its dimensions. Communication and harmonic living with others inform self-awareness and enable consciousness to uplift its capability. In the fulfillment of others we are fulfilled; and in the understanding of others we come to understand ourselves.

The way out of our present predicament is by means of the very agency that has caused us to become a cancer to other life. Sometimes cancers reach an impasse, which stymies their capacity for further aggression. When consciousness is fully informed in its malignancy, it may engage self-reflectivity to desist from enabling its destructive consumption. Personal and collective conscience is the informed state of self-reflection that can motivate from wasteful consumption to constructive sustainability.

The global shock of the terror of 9-11-01 might function as a catalyst to jolt the hyper-consumerism of Western cultures into a realization of the likely future outcomes of internecine, cannibalistic malignancy. Humankind must awaken to the causal elements of self-destructive violence as well as to the causes of life's sustainable success on Earth.

Humankind has learned what it means to be a "conscious" agency, but it has not yet learned what it means to be a "conscionable" agency. The development of conscience should evolve from *religious consciousness*, from the ethical awareness that preying mortally on others is an unconscionable rationality that leads ever more deeply into the wasteful malignancy of self-destruction. In our time, global economics are presently driven by the uneconomical motive of profiteering on global resources and of exploiting indigenous peoples.

Need, not greed, should be the motivating insight of true economics, which bases itself upon bioregional resources and bioregional needs. As part of a sustainable ethic, resource preservation and augmentation should occur within the bioregion even as the resources are being used. The sustainable exploitation of resources needs to be kept within limits that secure the economic/ecologic vitality of the "middle tree". The biblical story of the Garden of Eden and the "first sin" is a cautionary tale for the edification of all humankind, namely, that the wasteful consumption of the *middle tree*—bioregional vitality—ultimately leads to internecine destruction, the Cain/Abel syndrome. Cain's mortal terrorism against Abel warns of repetitions of cultural terrorism if greater intercultural communication and consideration do not take place. The resources of Earth are finite, and if finite limits are ignored, violent consequences will continue to be provoked and life will be a continuously bloody warfare.

Through communication new knowledge may come to be attenuated in mass consciousness. If the new knowledge corresponds fundamentally with nature it has the potential of radically challenging and changing old presumptions. With the mass attenuation of new knowledge, mass consciousness may come to a new threshold and may engage a critical percentage of the population in a *paradigm* shift of behavior. Arguably, such a paradigm shift has occurred in religious consciousness, away from fideism, from fundamentalism and the old creeds that generally are premised in a static-centrist cosmology, from the fideist theology of divine *being* disconnected from creation *becoming*, and from a dualistic philosophy that divides the realm of being (spirituality) from the realm of becoming (materiality). Evidence of such a paradigm shift is the global falling-off of public support for institutional religions that cling to old orthodoxies, to exclusionary sacrament, and to fideistic expectations of literal belief in scriptures.

The energy of mass consciousness is contagious; put in quantum terms the resonance of mass psychic energy is (may be) attenuated in the consciousness of individuals when they are exposed to it. The term "ecstatic resonance" has been used [Sylvester Steffen, "Primary Scripture, Cosmic Religion's First Lessons", 2001, pg 46, www.1stbooks.com] to describe a condition of harmonic attenuation wherein a transfer of energy (consciousness) causes a quantum leap from one level of energy (consciousness) to another. A quantum leap of consciousness results from the neural sharing of electrons in new arrangements. A new state of electron sharing by a critical mass may result in a sea change of consciousness.

Trimorphic resonance, the coherent energy of interactive rational mechanisms of expansive complexity and of universal subtlety, remains the accessible means for self-conscious humanity to self-accommodate one another in their naturally sustaining milieu. These trimorphic, transformational mechanisms of resonance are products and processes of the interactive agencies of communication, consciousness and conscience, the naturally developed processes of communal symbiosis, of personal, ecstatic resonance.

Patriarchy and Breach of Trust

In the catechism days of my childhood (1930s-40s) the notion of conscience that I acquired was very narrow and targeted, as was the sense of morality that constituted the palette of conscience. And I don't think the catechesis I received was that different from that of other Catholic kids of the time. As to the content of morality, it was mainly framed by the "thou-shalt-nots" of the Ten Commandments, more so than by the "thou-shalts" of Jesus' teaching. Making up a large component of the moral content were matters pertaining to sex, which on the scale of things seemed to dominate all others. The coloring of values with sexual overtones seemed to cloud all morality, distortions, which now seem rooted in the biases of institutional patriarchy. Women, their sexuality (and what more than that characterizes her personality) were seen, for example, from the moral perspective, as obstacles to spirituality and to moral altruism. As the primary agent, next to the devil snake, woman was the major temptress and agent of original sin and of the fallen-ness that still destines humankind with suffering. This overview was part and parcel of the catechesis of Vatican I theology. But, then along came Vatican II and an entirely different vision.

While "conscience" in Vatican II theology retains the meaning of "choice", informed *cum scientia* (with knowledge), the consciences of Catholics are not now held bondage to the service of institutional patriarchy, notwithstanding lingering biases to the contrary. The habituated fear/guilt impositions of old church on personal conscience are nothing less than breaches of trust. Growing distrust toward cultured patriarchy is variously manifested in society as well as in institutional Church. The falling off of Church attendance by the faithful and its disillusionment with the male celibate priesthood may be direct consequences of the distrust suffered from the long vogue of cultural patriarchy. If this is true, it is a sorry commentary on the state of institutional Catholicism at the present time. How to remedy this misfortune?

The institutional Church it self could initiate remedies by showing greater trust in lay people and by discontinuing its formalized practices of alienating women—a good beginning would be the reconsideration of universal priesthood and the confirmation of it by institutional priesthood. Clinging to discriminatory structures as practiced in Old Testament patriarchal culture is wholly inappropriate to Christic sensibility and to modern consciousness. The logic for change is bolstered in a theological understanding of paradigmatic nature that builds on the quantum-electric codependency of femaleness and maleness. A little reflection makes it apparent that the theology of patriarchal authoritarianism motivates institutional preference for male dominion in priesthood and Church hierarchy, a rationalized carry-over from Old Testament mentality.

What distinguishes the imaginative capacity of human rationality is its capacity to anticipate consequences of good and bad actions, and of true and false presumptions. This faculty of "prevision" empowers humans to apply correctives to their conduct, and to put in place assumptions and actions that have good consequences rather than bad. Thus, we may recognize the rationality of conscience as the "faculty of prevision". Conscience, motivated by love, is the rational faculty that sustains humankind personally and socially, and as such, it is the ultimate resource and virtue of "incarnational" success.

Incarnation is quintessentially a female process and *proclamation*! The alienation of women from the institutional ritual of eucharistic incarnation/proclamation comes off to thoughtful people as a form of rejection of the natural incarnation process itself, a kind of psychological abortion if you will, in the least a bias against femaleness, which flies in the face of Jesus' choosing to be fleshed in the body of Mary. Women's alienation is a counter-proclamation against the good news heralded by the mothers of Jesus and John the Baptist to each other, by their unborn babies to each other, and by the life and instruction of John and of Jesus to the apostles/disciples. The absence of women now from a celebratory role in the eucharistic proclamation/celebration of the Mass seems more a bad circumstance

147

of cultured male exclusivity, preferred in the priesthood of Old Testament theology, than anything divinely intended. Maleness and mandatory celibacy in the Christian priesthood seem on a par with the mandatory Judaic precondition of circumcision for Christian membership.

The present time alienation of women is a cultural statement of hyped male self-affirmation. Patriarchal elitism yet endorses the prideful *lucifers* of judgmentalism, authoritarianism and dogmatism. In contrast to the Trinitarian virtues of faith, hope and love, these isms are anti-prophetic demons that are without virtuous sustainability, either for individual or social personality. To the contrary, they are alienating forces of centrism that defeat the positive and sustaining energies of the theological virtues.

Because it breaches trust, patriarchal religion is antithetical to the Christian mandate to love God and one another. The mandate of the theological virtues is a mandate of trust, of covenantal mutuality. Faith, hope and love are emotionally sustained and sustaining powers that are reason-based. Faith flows from the trustworthiness of people toward each other, and is affirmed in their dialog and discovery of the mutually reinforcing powers of truth. Hope is evoked in mutual consciousness that flourishes in people from the experience of mutuality and truthful relationships. Love is the bond that causes mutuality and social well being to thrive. Love is the conscionable motive that convinces individual consciousness to exercise personal self-restraint for communal betterment.

Our faith tells us that truth will prevail. Where the cultural suppression of truth prevails, as in the cult of patriarchy, the virtues of rationality will gradually surface as the countervailing falsehoods come to be exposed. But change away from old habits of error doesn't happen easily for their beneficiaries are too accustomed to their benefits. The faith-affirming way to make it happen is to expose error to the light of truth. And who will expose patriarchy's sin and error? Certainly, the patriarchal institutions are not likely to admit that they are wrong. The weaning of societies away from male-assertive hyper-culture will have to be done by women, especially by women theologians, for they have particularly credible standing against the anciently churched and deeply rooted patriarchs.

The future of family and societal well-being depend on the tough-love of female rationality. It is right for women to cultivate in their children a faith-conscious sense of interdependency that says "no" to the hyper-culture of male advantage and exclusivity, whether in Church institutions or in secular society. Women should move forward in this motherly work knowing that truth will prevail and that they have the support of fair-minded men of faith and good will.

Schism of the Vatican Councils

Over interpretations of "revelation" and "prophecy", the Church of Vatican I and Vatican II is divided. Vatican I, the *revelation Church*, holds that after the apostolic time, direct, divine revelation occurs no more. Prophecy becomes an exclusive apostolic charism, reserved to the ordained (male) hierarchy. The *prophetic Church*, the "People" Church of Vatican II, holds that by virtue of baptism and inspiration of the Holy Spirit (divine consciousness), universal priesthood (prophecy) is a calling of service, revealed and revealing in the spiritual dialogue of persons with their indwelling God. What are people to make of this seemingly irreconcilable division? It seems appropriate to believe that the charisms of revelation and prophecy are alive and well today in *God's People Church*. People of this belief hope to mitigate the animus and incivility now frustrating Church credibility and its service to the people.

The theology of "closed" revelation underlies the closed, absolutist theology (and politics) of the Councils of Trent (1545-1563) and the First Vatican (1869-). Tridentine faith holds that God elected and chose the institutional Church of Roman Catholicism, in hierarchical structure and theology, to dispense salvation grace by and through its exclusively male priesthood in the celebration of ritual Sacraments, especially the Sacrifice of the Mass, established by Jesus Christ, Son of God and High Priest. (*This theology is consistent with the recent and controversial Dominus Jesus statement, authored by Joseph Cardinal Ratzinger, Prefect of the Congregation for the Doctrine of the Faith.*) The sanctioned paradigm of ritual sacrifice for the expiation of sins is deeply rooted in ancient Judaic (patriarchal) theology, and in the (male) priesthood of Melchisedek. It was realized already in the Old Testament that God asks not for blood sacrifice and sin offerings but obedience. The life obligation of personal, sacrificial giving is an obligation of everyone and is not the purview of a class of elites.

The doctrine of faith, including original sin, as developed by Trent and Vatican I, sees the dispensation of grace and salvation, as flowing from God-Jesus, through Church hierarchy, priests, to the people. Evolved from this theology of priesthood was the corrupt practice of the Church in selling *indulgences*—merited grace—to the laity. Protesting priests of good conscience demanded reformation of this egregious breach of faith and abuse of authority. The promise of indulgences for the remission of sins was an act of extortion for it succeeded in collecting from the faithful great sums of money that built Churches and monasteries, and financed Church-sponsored wars.

At the deep heart of ancient Judaic theology, also developed in Christian theology, is *God's expectation,* a people-grown consciousness—of peoples' behavior in their relationships with each other and with creation. Questions needing to be answered today are, "Are people called to be submissive and passive to the institutional 'norms' put forth by the Church leadership of another time claiming absolute right with respect to the dispensation of grace?" And, "Are people to function as bearers of grace to each other in Church institutions and under auspices of Church or in the venue of daily life?" These questions especially bother women, who, for the only reason that they are females, are excluded from Church hierarchy and from "ordination in the dispensation of grace". Many feel that male priestly exclusionism unjustifiably alienates women from God's salvation plan and with bad consequences for God's whole family.

The recent experience of a young woman Church activist, Ms. Anice Schervish, was published in the Winter 2000 issue of "New Woman, New Church", the newsletter of the Women's Ordination Conference (www.womensordination.org). It was reprinted in the January-February 2001 issue of "ChurchWatch", pg., 4, the bi-monthly newsletter of Call To Action, 4419 N. Kedzie, Chicago, Illinois, 60625. Ms. Schervish arranged for a face-to-face discussion with Cardinal George, of Chicago, about women's ordination. Predictably, the meeting was inconclusive. However, the point of theological departure was sharp. Cardinal George instructed her "one's own experience is not, and cannot be, 'normative' for the Church." One might agree, however, that universal, *consensus experience* of the faithful can be normative. Consensus experience corresponds with the "sensus fidelium", which should be normative for the Church. Ms. Schervish writes, *"I pointed out that those who have been determining the norms for 2000 years have all been celibate males. I challenged him to use his position to be prophetic. He said that prophecy is standing for Truth that is Jesus Christ. I said that we are being prophetic. He disagreed, insisting, that it's not about experience but about revelation. He said that we are being influenced by culture. He said [that] we are not an enlightenment people, we are a people of revelation, which was closed with the death of the last apostle. I personally find that hard to believe, and I told him so. He wasn't amused."* Theological tension over *revelation* and *prophecy* is showing more frequently and in diverse manners. Interpretations of *collegial,* theological authority are in conflict. The excommunication of Nebraskan "Call to Action" Catholics by a local bishop is an exercise of Vatican I Episcopal arrogation.

The "reformation" of Roman Catholicism as enacted by the Councils of Trent and the First Vatican is one public face; the other is that both Councils were calculated to counteract the Protestant Reformation. The transparent truth is that both Councils were power initiatives of institutional Roman Catholicism, intending to stanch the successes of Protestantism. Except for the Protestant Reformation, the Councils would never have happened in the way that they did. While it may be an oversimplification and somewhat gratuitous, it is not incorrect to characterize the "Reformation Church" (Protestant) as the *Prophetic Church,* and the Catholic "Church of Apostolic Revelation" as the *Counter-Reformation Church.*

Robert Cardinal Bellarmine, S.J., (1542-1621), "distinguished himself in the Council of Trent and as principal consultor in the Roman Inquisition trials of Bruno and Galileo". [Ramon G. Mendoza, Ph.D., "The Acentric Labyrinth: Giordano Bruno's Prelude to Contemporary Cosmology", *Glossary,* Bellarmine, 1995, Element Books,

Inc., P.O. Box 830, Rockport, MA 01966]. It is an historical fact that "worldview" has everything to do with theology, especially at the Reformation time. Bellarmine knew this, and so advised the pope, (Id. pg. 37). Insight with the God of creation is qualified by the human understanding of creation, the venue of ongoing *revelation*.

The worldview of the Councils of Trent and the First Vatican was earth/man-centered—the static/centrist view of Aristotle. To the Church it was clear that except it confronts vigorously the revolutionary worldviews of Copernicus, Bruno, Galileo, et al., they could become a tool that would give impetus to the Protestant Reformation. Cardinal Bellarmine warned Pope Clement VIII of this high risk. It was accordingly Bruno's unorthodox views of Earth and the universe that motivated the Roman Inquisition to condemn him and to burn him at the stake (1600). Cosmological positions that got Bruno in (theological) trouble include: that the cosmos is centered neither on Earth nor on the Sun; that the universe is infinite; that all substance, in *form* and *matter,* is a unity (homogeneity and isotropism); that the human soul is in harmony with nature; that the fluidity of nature isn't quantitatively limited.

Religious pluralism is here to stay as is female/male egalitarianism, and so is the fluidity of conscious transformation. Monarchy/patriarchy, the politics of dominion, whether by Church or State, is counter-intuitive and politically unacceptable. The dominion echoes of the past resonate only in heads fixed in the past. Moral consciousness of changing, real-world relationships is increasingly more compelling than dominion echoes from the past. Fixated religion is eschewed by an ever growing public. Faith, in dogma out-of-time, is couched in words no longer used, in phrases no longer believed. Life/faith is script for the future, even as hope is confidence in the unseen, so, personal conscience (intentional consciousness) looks to the future. The universal time/space-accommodator is communication, the informational agency of consciousness.

Consciousness refuses to be fixated in time. The consciousness of consequences prods conscience to reconcile consequences with actions, in real time. Inherent trust, faith, is birthed and augmented by the trimorphic process of rationality, individual and communal, and not by the fiat of self-affirming institutions fixated in out-of-time orthodoxies. The static mind of Vatican I is fixated in centrism and absolutism; the transformational mind of Vatican II on the other hand is open to acentricity and relativity; the former is conformed to an earth/human-centered cosmology/theology; the latter seeks to be informed in evolving cosmology/theology. To Vatican I theology, revelation is fixated, and is not an ongoing process, but to Vatican II, theology/revelation is open. In practice, Catholic Theology seems yet to stand outside the quantum-cosmic consensus of modern consciousness.

Vatican II seriously engaged *updating* by which the Churches of prophecy and revelation might be reconciled. The adamant choice of old expediency, if defiant of truth, will turn against its agents. New consciousness is in *spirit and truth*; being "born again" doesn't require returning to the womb, and neither is new consciousness enwombed by institutional constraints that defy the *new garments* of truth (worldview). The *marriage feast* is a continuous and open event. But a closed worldview does not gain entrance. All are invited, but few enter. Not only does scripture call us to cloth ourselves in new awareness, so does Vatican II. *"The human race has passed from a rather static concept of reality to a more dynamic, evolutionary one. In consequence, there has arisen a new series of problems, a series as important as can be, calling for new efforts of analysis and synthesis."* (Gaudium et Spes, Introductory statement, No 5, par 4. Emphasis added.) The closed theology of the First Vatican Council is a "temple theology" like that of Nicodemus, who couldn't bring himself to witness the public priesthood/discipleship Jesus invited. What ever became of Nicodemus? History seems to have entombed him in temple-cultured dependency, not yet ready for Jesus' priesthood that replaces the priestly patriarchal class of antiquity.

Lent and Relent
(Stones also Change)

Since humans are social, communal creatures, they do things together out of necessity. This probably explains the human tendency to institutionalize everything that can be institutionalized. When institutions are created they are done so, not in isolation from, but in context of the times, the social conditions, worldviews, presumptions and world values. Service institutions, like churches and schools, are flavored by political, economic and faith presumptions, which build on presumptions of ancient times and on simplistic worldviews from which moralistic mythologies are woven. The tendency of church institutions, developed in the context of the prevailing world order, has until now been to advance theologies that have their origin in mythical worldviews and moral stories and with an expectation of orthodoxy (literal belief in the handed down stories) that accepts the validity of storied details, without questioning theologies against the cumulative and changing insights of communal knowledge. It shouldn't be surprising that dated theologies, institutionalized in a whole different world context, become outdated and lose credence, even as do their institutions. Most modern religions, in some way or other, are confronted by the ghosts of the imagined past; this is true also of Christian religions.

Jesus made it clear that his kingdom was spiritual, not political and juridical, as earthly kingdoms are understood to be. Jesus came to announce a new historical era for people—all people—and religion, a new flowering of human consciousness. Up until his time, patriarchy prevailed; but Jesus intended to break with the old theology and worldview, and to begin the era of the messiah, an era of mutuality, of God entering and being associated with human consciousness in a new and entirely different spiritual perspective, a perspective of love prevailing in all person-to-person relationships. The old law scarcely glimpsed toward this ideal in that it was unable to break from its harsh conventions and from seeing religion as a person-to-God relationship. The reign of contrived ritualistic religion and of legalistic authoritarianism, dominated by the male culture of political violence and religious disregard for the marginalized, was in place, but was to be replaced with a new era of nonviolence and egalitarian concern.

Modern Christianity is still plagued with the violent vestiges of institutional patriarchy that came before it and that were to be replaced by it; patriarchy perhaps still prevails for the reason that the same worldview that birthed it is still in place in the traditions and theological underpinnings of contemporary Christianity. The radical theology of love announced by Jesus, the Word-Made-Flesh, hasn't yet replaced pre-Christian patriarchy. The ego-centrism of ancient patriarchy and its idolatrous obsessions still reign.

Wars of denominational pretense amongst the Christian religions have been and continue to be waged, and in their competitive endeavors, each to create its own worldly kingdom, the old egoisms of patriarchy again surface in faces like those of Old Testament times. Eventually, pretenses run their courses and come to be exposed for what they are. The gravity of the present times seems to contribute to a sense of urgent need for the resolution of the pretenses that have occasioned the many schools of denominationalism. The public no longer buys in to the artifices of denominational divisions, and like waves pounding the shoreline rocks, people press their institutions to change.

People individually, more so than institutions, are sometimes forced to change or to perish. It is true that institutions can weather storms that individuals may not weather. New circumstances may require new knowledge and new ways of dealing with them. Because of their size and resources, institutions may see themselves as bulwarks against the tides of change. But even boulders on the seashore change under the relentless pounding of the waves. Ever so gradually, their harsh edges are ground down; a little at a time, they are reduced to sand and serve to soften the crash of the waves. And so it is with institutions; those that adjust and accommodate to change survive, whereas, those that cannot accommodate become alienated from the people and lose their capacity to serve by losing public support. They are faced with a slow dying unless they embrace a radical updating that connects them to the circumstances of new times, free of old and meaningless fixations. A discernment of spirits may reveal when it is a time for them to relent from hanging on to the old and to adjust to the consciousness of the new. The spirit must ever be renewed if faith would be.

Causes of conscience are birthed in personal consciousness; they root equally in reason and faith. Noble causes, altruistically and humbly begun, experience frustration when institutionalized. "Why?" you ask, because corporate profits are the primary concern of corporate expectation. Church tames saints and elevates sinners by canonizing them. By definition, counter-cultural causes (with which saints are often associated) are counter-institutional. Jesus is example *par excellence* of counter-cultural conscience for his mission uniquely joins the causes of the culturally alienated and disadvantaged. The flip comment is made with justification that the sure way to frustrate a worthy cause is to institutionalize it; to the extent that institutions exploit for damaging profits, they prostitute. Institutionalizing always involves bureaucratizing, and bureaucratizing always means structural expense, expense to hire people and to define and promote the mission and work of the institution. The principle of subsidiarity is a principle of economy, which says that every product or service should be provided as directly as possible at point of need and from direct sources—not through some complex bureaucratic chain layered in commands of personnel, which burden excessively with costs layered on products and services. [APPENDIX B. White House Conference. Pg. 271]

There are legitimate goods and services that are better provided by people working together, as in institutions, than by people individually; and, people need to be paid for their services. But, the wasteful loading of goods and services with layers of personnel leads to inflationary pressures that burden people as well as resources. The public trading of goods and services in paper transactions (stock market), because it pays people not directly involved in producing goods or in giving services, can be economically and ecologically wasteful. These paper markets are motivated solely for the profit of stockholders without regard for the validity of the product or the merit of so-called "services". Paper transactions, motivated solely in profit, invite scams and the "marketing" of products and services that have little or no real value.

Traditional theories of economics continue to ignore the real values of natural resources and ecologies, and consequently, they rationalize and enable the exploitation of resources in ways that are insensitive to wasteful profiteering in terms of the entropic deficits inflicted on natural resources and ecologies. The neglect to include real ecological values in the exploitation formulae of economics allows for the layering of profits on natural resources, such as minerals, lumber and fish, and therefore causes redundancies of cost pressures on existing resources, which overwhelm their capacity for sustainable recovery.

The institutions of government, education and religion are institutions that "market" services to the public. The costs for their services are borne directly by the public. In the service work of these institutions, products are needed to support the service, such as reading materials, which educate, explain, edify and exemplify. This need gives rise to the publishing trade and to adjunct institutions. Consider, for example, all the religious publications used and promoted by churches, not to mention the throwaway materials required annually by educational institutions. The costs of these goods and services fall on the public and transfer ultimately on to the ecology, which provides material for paper, ink, presses, etc. Market driven economies, which depend on the production and marketing of goods and services to the public, become ever more taxing on ecological resources as the need for goods and services grows. Theories of market economics presume a limitless resource of raw materials from natural ecologies. This presumption is fatally flawed because the resources of nature and nature's capacity to sustain the pollution of manufacturing are not unlimited. The cumulative damage of resource consumption and environmental pollution is causing the shutdown of interdependent life networks and the acceleration of species' extinctions. When corporate profit rather than public welfare is the motivation driving corporations, the consumptive cost to nature becomes ever more taxing.

Historically, as part of the moral contract (covenant) with nature, people recognized the burden of human subsistence on nature and their need to facilitate nature in healing her deficits. In Old Testament Scripture a religious strategy is provided for the recuperation of both humans (Sabbath) and nature (Jubilee). Discernment of the natural place of humans in general and of the individual in particular has always been part and parcel of social ritual and culture. This was also true for Jesus as he became ready to enter into his public life. In our time, Christian religions celebrate Jesus' discernment trial, which he undertook and which clarified for him and for us who commit to be his imitators the options in life that should be preferred.

The forty-day trial of Jesus is referred to in New Testament Scripture as his "temptation". Christian Churches have instituted in their liturgies the celebration of Jesus' forty-day trial; it is the period of forty days in the Spring-cycle of the Church called Lent. Personal focus is directed on the need for self-denial, fasting, good works and alms giving. As is the case with so many things that are celebrated in calendars of ritual, focus on them is only in the timeframe of the cycle. Jesus trial is a life-determining event, not just a Church event intended only for the forty days of focused attention to it. Attention to the forty-days spiritual discernment that gave focus to Jesus' life should also give focus to ours, not only for forty days of the calendar year, but for 365 days. The institutional calendarizing of this event is a case in point of how the institutionalizing of something can serve to remove it from the purview of daily living and make it a *religious* event rather than a practical, *secular* one. Jesus' life discernment on the occasion of his "temptation" was decisive in the way of his "secular" living, which was one and the same as his "religious" living.

Jesus weighed aligning himself with the Church institution (temple service) and/or with the public exploitation of natural resources for personal gain. His option was neither, rather he opted one-on-one service to the needy. New Testament Scriptures document how Jesus lived out his life-decision not to be institutionally aligned, but to serve personally and altruistically, and to live communally, sharing with others both good fortune and bad.

The institutional marketing of goods and services includes good and bad elements, depending on the motives of people making up the institutional bureaucracy; elements include motives of altruism and obsession. When motives of obsession are stronger than motives of altruism, institutions, like the people within them, become predatory. Jesus railed against the predatory laws of Rome and the burdensome rituals of the Temple, and he committed himself to liberate the people from both.

In our times, institutional motivations for corporate profiteering play on public obsessions. Notably, public trading (stock markets) on the profit potential of corporations that "go public" has become a risk game that plays on gambling instincts. Shocking scandals of fraud and profiteering, [e.g., Enron], reveal fundamental flaws in governmentally legitimized devices that enable public predation. Enmeshed in the sticky webs of public predation are publicly elected representatives who are double agents, purportedly serving the public good but serving corporations when it serves them personally, even if it is against public welfare. When democratically elected officials serve corporate obsessive-ness over public well-being, government becomes perhaps the most pernicious player in the public game of defrauding. The mutually reinforcing motives reinforcing the unholy alliances of government and corporations are snares that entrap people and resources in webs of waste and violence. Public order and welfare need public institutions, but needed also is oversight that prevents betrayals of the public by bureaucratic collusion and greed.

The crisis of the present time is waste, pollution and exploitation caused by heavy living and resource devastation. The sustainable welfare of all life requires restrained, conservative usage of resources, not their unbridled consumption in the service of excessive self-aggrandizement.

Where is religion? Has it become but another contemporary institution not above the game of preying on the public? Before Lent can become a life-determining event, individual and institutional consciences need to relent from their habituated misdirection and become conscionable agents of public service. People must become knowledgeable enough to know what works, which services and goods are altruistic and which play into the category of predation and resource waste. Only then will we truly be conscionable in our living and will our religious and secular living comport with each other.

Dimensions of Cosmology/Theology
(Planes of Consciousness)

The theological dimensions of relationships, from the perspective of quantum relativity, have quite an important difference in emphasis than those from the centric cosmology perspective. Perspective, like perception, is an aspect of consciousness. So, with respect to theological perspectives, whether in the perception of quantum relativity or centric cosmology, the descriptions of the dimensions of relationships fall under the common heading "planes of consciousness".

In the perspective of traditional Christian Theology, there are two planes of relationship. The first plane has to do with the human-to-God/God-to-human relationships. The second has to do with human-to-human and human-to-other life relationships. The first is called "transcendent", and the second is called "immanent".

The term *transcendent* pertains to knowledge that is beyond the limits of human experience, prior to and different from (embodied) human relationships. Such knowledge exceeds and surpasses all knowledge derived from sensation. The plane of this knowledge, in traditional theological language is "vertical".

The term *immanent* pertains to knowledge that comes to human consciousness through the senses. As the old adage has it: "nothing in mind except first in the senses". This knowledge is mind-embodied knowledge. The plane of embodied knowledge, relationship, is called in traditional theological language "horizontal". Charles Dickinson describes these planes very succinctly, "the vertical is the transcendent God-man relationship, whereas, the horizontal dimension is the immanent, earthly man-man relationship". ("The Dialectical Development of Doctrine", pg 303.)

A nearly equivalent term that might be used for the vertical plane is "from above", while an equivalent term for the horizontal might be "from below". It is apparent that the singular dimension of each plane is simplistically linear, and that the analogies limp very badly with respect to the true complexity of relationships. The dimension of the "essential continuity" of energy/matter, in time/space terms, is not at all dealt with in the centric cosmological paradigm of traditional theology. The addition of one more term might serve to bridge the gap and give to static-centrist cosmology/theology a way of joining the cosmic dimensions of energy/matter/time/space—and may apply to *Christology*, the theology of Christ Present. The word I offer is "imminent". The word *imminent* means immediate, impending, at hand, about to occur. The term ascribes morality to the qualifications of *time,* and to its quantum correlative *space,* in addition to the linear vertical and horizontal dimensions.

How may the theological perspective on imminence in quantum relativity change the human/Earth-centered perspective? The meanings of vertical and horizontal are simple enough to understand, linear relationships. The physical disposition of Earth in relation to the individual person is horizontal. Experience and conventional usage of the terms tell us that this is so. Individually, each of us is located at the intersection of the vertical and the horizontal; when we look vertically up, we look into the realm of divinity; when we look around us, we scan a circular horizon that defines that part of Earth, which falls within the scope of our vision. Our scope of Earth locates the source of our material needs, which are derived from Earth. We observe that earthly vitality is a benefice that depends totally on *vertical* graces, sunshine and water. Grace, life itself, ultimately comes "from above", from the place (space) of the transcendent. All individual life on Earth is short term, and life's grace quality, soul, is destined to return to the transcendent, whence it originate(s)d, the abode of the Deity. In the abode of the transcendent, neither time nor space exist, only pure spirit which is without dimension and whose spirituality isn't tested by moral valuation. In Christology, "from above" pertains to divine nature; "from below" pertains to human nature.

In material existence, in real life on Earth, time and space are essential qualifications of relationship, which bear directly upon the moral, ethical conduct of creatures in their relationships with each other. In the theological analogy, speculation with respect to spiritual, non-material existence derives its perspective from experiential consciousness (sense) of Earth/human/other relationships. By analogy, Earth/life experience is extrapolated onto extraterrestrial (transcendent) spirituality, the realm of theology's "supernatural".

The fact of (the) matter is that ethics, morality, are (theological) issues in the immanent plain of Earthly existence. The *time plane* is in three dimensions, which are essentially continuous, codified in conscious flesh, and transitional into each other; the three dimensions of the time plane are the past, the present and the future.

The present time coding of the past is a qualification of DNA, genetically transmitted to each cell; present time experience is implicated genetically in the forms of psychically qualified packets called *memes* (Dawkins). Genes pertain to physical, natural, physiological expressions, and memes pertain to psychical, nurtural, cultural expressions.

The physical necessities of life are finite, that is, they are limited by mutual interactivities of diverse life forms making up symbiotic web systems. It is in the nature of living relationships that all life forms give origin to each other and depend on each other for individual and communal sustainability. Morality and ethics enter into the domain of "right to life" and pertain to the mutuality of essentially interdependent life forms. Instincts guide the inter-relational activities of non-self-aware life forms, but humans, because of their intellectual and judgmental capacities and liberties can exploit resources to extinction. The sustainability of all life depends on the health and sustainability of web systems and on the symbiotic balance of all codependents in their relationships within their webs. Thus, the securing of communal health and sustainability, the balance of ecological interdependencies, seem to be of a higher moral order than the individual right in claiming against network and communal sustainability. Except for communal sustainability, individuality has no existence.

Which is the greater moral wrong? The destruction of webs of life that make individual life possible and sustainable, or the right of individuals, one species, to hyper extend its individualistic claims and cause the collapse of the web? The proportionality of the moral dilemma seems clear. The present vertical-horizontal dimension of moral relationship in centrist theology fails totally to valuate the time/space moral dilemma. Unless theology addresses the terminal threats to life that come from the prolixity of human presence on Earth, and the claim of individualistic right over communal, it fails in its moral obligation to valuate the moral proportionally of the God-man-other-life codependency. This issue seems to be totally absent from the moral dialogue occurring in theology, except for the "this world" focus given in Liberation Theology.

What moral obligation does contemporary humanity have to future humanity? Are we not ethically obliged to preserve a sustainable and sustaining ecology? If we fail in this, we put ourselves at risk of aborting the sustainable future of communal humanity. This moral issue is over and above the moral issue of the right to life of all symbionts interacting within the interactive webs of life. Isn't the issue of restraining the excessive prolixity of one species within the web and the securing of the web's sustainable health of a higher moral proportionality than insistence on individual consumer right and risk of the collapse of the web? Rights of web members and the preservation of vital balance are above all the protective moral obligation of humankind who alone threatens the webs and who alone has the free-choice judgment to stop its cannibalism.

The pretended theological ignorance in the matter of crass consumerism is an especially heavy sin on institutional religion for it culpably sanctions nature's desecration even as it profiteers on her mortal misfortune. Religion/theology that fails to recognize the proportional morality of web sustainability over individual is arrogantly fixated in fatally flawed linear thinking. If theology fails to process beyond its linear dimensionality it will be judged as dysfunctional in the quantum world reality, and public consciousness will ignore it, and move on in recognition of the moral implications of quantum relativity, and make judgments on its own that fly in the face of uninformed theology.

Health and Healing

Harsanyi and Hutton ("Genetic Prophecy: Beyond the Double Helix", 1981, Rawson, Wade Publishers, Inc. NY, pp 10 & 11) make the blanket observation that *predisposition* and *insult* are always co-factors in the etiology of illness. Physical disorder and psychological disease arise from mind/body predisposition, and/or mind/body insult. Factors of predisposition (genetic, environmental) and insult (genetic, environmental) together work havoc on health and healing. Obviously, if this is true, secured health and healing involve correction of disease predisposition and avoidance of environmental insult. The combined havoc of external/internal factors of predisposition and insult attacks and breaks down the mind/body defenses of natural immunities. Through mind/body communication entrance of the external factors of predisposition and insult is gained into the human system. Recognizing and avoiding communications that introduce insult and predisposition to disease into the mind/body, and living habitually in a manner that makes for a healthy disposition of mind/body, should be priority concerns of all who value health of mind/body. Healthy *soul-food* (intellectual nourishment) and healthy body-food (body nourishment) make for soundness of soul and body; unhealthy soul-food (trash reading) and unhealthy body-food (junk food) make for soul/body predisposition to disease and disorder.

Global industrialization, depletions of rainforests, environmental poisoning, mass starvations, and over consumption of drugs and over processed foods are agencies contributing to disease predisposition and insult. Catastrophic failures of immune systems seem to be a new and lethal thrust of modern day apocalyptic horsemen.

Health and healing are issues of religion (right relationships) and of civility (socially secured life necessities). Health and healing are spiritual/physical conditions of sound mind in sound body, which are the purview of religious sacrament ("worthy purpose") and civil accommodation (enabled access to life necessities).

Health and healing are consequences of intended, *worthy purpose*, whereas, sickness and disease are consequences of purpose not worthily attended. Failed attention to worthy purpose in matters of personal behavior is an agency that wastes immunities and incline body systems to the inner dissolutions of health-guards by the invasions of outside agents of insult. Dissolute habit exacts a tax against health and well-being, and in ancient times, the consequences of dissolute living were seen in the religious perspective as punishments from God. Whether nature's judgment or retribution of divine justice, the man-made causes of the bad consequences of human disease are the same and are wisely avoided whenever possible. Mindless bad behavior and intended bad behavior may have the same bad effects; mindlessness and bad disposition both need to be confronted by personal good sense—conscience—in the interest of well-being.

Sexuality is purposeful and beneficial to health when purposefully engaged, but not helpful to good health if engaged mindlessly and in self-serving, abusive intent. When mindless and intended misconduct are opted, whether personal (sexual misconduct) or social (environmental pollution), one's own person, as well as other's, may be exposed unnecessarily to disease predisposition, and, not God nor nature can protect against the consequences of habituated actions of abuse. Disposition habituated in insulting behavior hastens the day of disease. The global disease of Acquired Immune Deficiency Syndrome (A.I.D.S.) might profitably be weighed against the foregoing observations.

A.I.D.S. and Sexual Promiscuity
10/21/85. SLS.

Pestilence and starvation are creature realities that predate human consciousness; consciousness of their counterparts, health and nutrition, is of more recent origin. During the 1850s and 1860s, Rudolf Virchow was advancing his cellular explanation of disease in Europe. Before the turn of the 20th century, studies of skeletal and mummified tissue were pursued to correlate ancient diseases with modern. This pursuit has been greatly advanced with the refinement of biological knowledge and with the perfecting of laboratory technique. The U.S. Anatomist Roy L. Moodie is one of the first practitioners to extend his discipline to virtually the entire biological realm. His text "Paleopathy: An Introduction to the Study of Ancient Evidences of Disease" (1923) is a classic that includes diseases of virtually all plants and animals. Obviously, ancient disease is deduced from evidence persisting to the present time in preserved remains of ancients.

Determining specific organisms that are responsible for specific diseases is quite another problem. Even diagnosing causes of modern day disease is difficult because the responsible organism(s) may be altered or destroyed in the disease cycle. Only a few ancient diseases leave traces distinctive enough for diagnosis, as for example, degenerative bone and joint disease. However, in ancient writings, including the Old Testament, significant references to disease are found. (Kubet Luchterhand, "Health of the Ancients", 1982

YEARBOOK OF SCIENCE AND THE FUTURE, 1981, Encyclopedia Britannica, Inc., pp134-147.)

When one is stricken with disease it is quite natural to wonder what caused it. If one is infected with a tumor one wonders if it is benign or malignant, and what causes it to be one or the other. Bacteria and viruses are known culprits causing disease, but often little is known about their origin and what it is that makes them benign or life threatening. One questions how bacteria and viruses fit into the evolutionary scheme, how at one time they may behave symbiotically and at another time be catastrophically hostile. It is indeed quite mystifying to realize that the cells of all living organisms, including human, exist because of symbiotic alliances of untold numbers of onetime independent organisms, bacteria and viruses, which, in the course of untold numbers of years, have tamed their hostilities toward each other and have come to a remarkably harmonious accommodation, and even interdependency, a common "sacra mens".

[2/10/03. About bacteria, E.O. Wilson observes: "...Different strains {of bacteria} and even species readily exchange genes, especially during periods of food shortage and other forms of environmental stress. Their generations are extremely short, allowing natural selection to act on new assortments of genes within days or even hours, shifting the heredity this way or that, perhaps creating new species." ("The Diversity of Life", 1992, Belknap Press of Harvard University Press, Cambridge, Massachusetts, pg 145)].

Every living body is not really an individual entity, but a community of many millions of microorganisms functioning in symbiotic harmony and presided over by the brain. The same can be said even of the individual cells of bodies, namely, that they too are made up of a mix of once independent organisms evolved to an accommodation of interdependency. (Lewis Thomas, "The Lives of the Cell", 1974, The Viking Press, NY.) And what is even more astounding is the ability of individual cells to divide and to give rise to two cells, which are in every way identical.

Cytoplasm, the composite mixture within cells, [including eggs], is [a continuity medium] of maternal origin. Stephen S. Hall, ("The Fate of the Egg", SCIENCE 85, November, Vol. 6, No. 9, The

American Association for the Advancement of Science, 1333 H Street, Washington, DC 20005, pp 40-49), develops the fertilization saga of the egg (of XENOPUS LAEVIS, the African clawed toad), and the destiny of the zygote as being very much a geographical phenomenon, that is, a phenomenon of the egg's geographical surface shifting in a specific way that puts hemispheric egg contents in contact. The egg may be considered analogously with the Earth, that is, geographically differentiated. The egg has northern and southern hemispheres, with a North and a South Pole, and an equator between them. The egg is like other cells in that it carries the symbiotic mix of cytoplasm, but unlike other cells, it carries only ½ of the genetic DNA needed to develop a new toad. All materials contributing to the structural formation and functioning of the toad embryo are geographically disposed in the egg.

The two hemispheres of the egg have each their own specific destiny in embryo development. For example, the northern hemisphere gives rise to animal tissue and function, while the southern hemisphere gives rise to vegetal tissue. Magnetism and polarity work in the egg-globe in elemental ways. It so happens that the sperm can penetrate this egg only in its northern hemisphere, and when it does, tremors are set up which cause the outer crust to shift. Egg tectonics. The shifting puts materials of the northern hemisphere in contact with materials of the southern hemisphere, and the mix there (in the southern hemisphere) develops to be the dorsal region of the animal, which in all vertebrates becomes the spinal chord and brain. If the shifting of the egg crust is anything less than 30 degrees, the embryo will not develop. The successful engineering of the early shifting of inner egg-magma is critical to setting in motion the gene code sequences which program the lifetime processes of the new organism.

In sexual reproduction the whole point of the male contribution of ½ the nuclear genetic component is to mix the genes, that is, to create "inherent uncertainty" as to the genetic make-up of the zygote and the eventually mature individual.

Even in life forms that develop from bi-sexual reproduction, as in humans, cell cytoplasm is wholly female and has evolved originally and continuously from asexual reproduction, that is, without the contribution of males. Within the cell cytoplasm are "organelles", some of which, even today, preserve their ancient female originality and their own DNA. (Source: L. H. Snyder and P.R. David, "Principles of Heredity", D.C. Heath and Company, Boston, MA).

Mitochondria and plastids are two such organelles, which have their own DNA; and whence come their DNA? The ancient precursors of mitochondria are bacteria; and the plastids, like the chlorophyll of algae, may also derive from a common precursor. In the organism Hydra, a primeval example of animal and vegetative symbiosis persists to the present time. Chlorophyll produces food (sugars) by photosynthesis, and plastids play a role in synthesis of ATP. Mitochondria are microbial, rod like structures within cells and are responsible for deriving aerobically the energy from ATP. Mitochondria are very much like free-living bacteria, and have characteristics more in common with bacteria than with the cytoplasm of the cell in which they exist and from which they are structured. Because of animal and vegetal symbiosis in Hydra it deserves more attention.

Hydra has successfully incorporated chlorophyll from green algae, which provides food for it. When Hydra cells divide, grains of chlorophyll are equally divided between the two ensuing cells. In sexual reproduction, the ovum of the Hydra is the carrier of the chlorophyll grains, which transfer to the new Hydra. Plastids, being chlorophyll related, are vegetative in origin. The Hydra and humans, indeed all organisms, are beneficiaries of ancient accommodations of vegetative (plant) and microbial (bacterial) organisms whose individuality is suppressed today by symbiotic accommodation. (Lynn Margulis, "Symbiosis and the Evolution of the Cell", 1982 YEARBOOK OF SCIENCE AND THE FUTURE, 1981, Encyclopedia Britannica, pp 104-121.)

Plastids in the endosperm of corn grain and plastids in human cells, for example, may derive from the same ancient precursors as the chlorophyll in the Hydra.

For untold centuries humans have speculated on the origins of living organisms. One theory, which persisted for a long time, was the theory of spontaneous generation, which held that certain kinds of organisms arose from others, or from decaying matter. Since the research of Louis Pasteur, this theory has been reduced to all but ridicule. However, the total discrediting of it may be premature. In a modified form, this theory may shed light on the sudden and unexpected appearance of epidemics; a case in point is A.I.D.S., Acquired Immune Deficiency Syndrome. Perhaps there is validity to the theory to the extent that symbionts in cells, due to control of symbiotic mechanisms, may revert back to a non-symbiotic condition and become themselves infectious agents when the control mechanisms are compromised. The symbionts in cells, and in DNA of cells, customarily function in orchestrated harmony with each other and give creatures a marvelous potential for diversification; in response to some catastrophic foul-up, these same symbionts may be thrown into confusion and to misbehavior. For example, a hostile enzyme, a hormone or other might originate from the chromosomes of cells in response to some powerful and adversary stimulus, like from some toxic industrial solvent. The virulence of such toxic havoc might cause an ordinarily benign symbiont to revert back to hostile behavior, and/or may trigger mutations. The spectacular success of evolution lies precisely in the capacity of organisms to play subsidiary and supporting, though different, roles with one another. Accidental reversions, exceptions to the rules, are surely to be expected because of the *inherent uncertainty* of particulate chemistry at the sub-atomic, atomic and molecular levels. Victims of A.I.D.S. may be victims of nature's structured ambiguity that surfaces under conditions of predisposition and insult, and that lose their usually reliable harmony because of predisposition and insult.

A potentially rich field of science may be in the area of identifying symbionts and micro immune agencies that secure the working of symbiotic harmony. Constant and intricate communication in many forms must be occurring in cells and between cells for securing soundness (justification) in the organism; not the least of which are immune systems, which seek out invaders and do away with them before they do serious damage. Perhaps the perniciousness of A.I.D.S. is in its ability to deceive the immune system to accept the retrovirus precisely because it is a recognized citizen of the system. If the A.I.D.S. retrovirus originates from the body system, this might explain why recovered retroviruses are different. Also, if the DNA of the retrovirus links up with DNA of the body, its at-home-ness in the human body and possible origin from it are evident. Medical science has concluded this much, that the retrovirus is transmitted by exchange of body fluids, particularly through sexual contacts, homosexual and heterosexual.

Human biology is linked to human psychology and morality. Knowledge of what benefits the body system and what harms it compel judgment to opt doing those things that benefit and to avoid those that do harm. In this matter, knowledge gives rise to conscience and moral judgment that opts for "good biology". Good theology pertains to good biology. Authentic spirituality and moral conduct arise from informed intelligence and knowledge of the natural laws that govern well-being.

Awareness of biological catastrophes is as old as human intelligence, whereas, pursuing knowledge of causes is new. Ancient people and modern fundamentalists are inclined toward obvious, literal answers. Catastrophes are seen as punishment from God; and with respect to sexual conduct, certain taboos were put in place, and in time became culturally imprinted in public consciousness—taboos intended to avoid divine retribution. Actions that had biologically harmful outcomes (signs of divine retribution, to ancients) were taboo and were labeled as sin. Biology and morality coincide in judgment of "sin".

While the language and reasoning of ancient wisdom differ in perception from the modern, they converge in intent, namely, that biologically hostile acts are also immoral acts and are to be avoided by sanction of public order; damaging actions are "unnatural" (contrary to good biology) and "sinful" (opposed to right reason—conscience).

Historically, ancient wisdom has been handed down in religious tradition and in cultural morality (laws). Cultural morality is sounded in social experience. Taboos against sexual promiscuity and homosexuality originate in the experience of pestilence associated with them. For example, personal morality of Old Testament Judaism came to cherish chastity and to hold the adjuncts of pagan worship (prostitution) in contempt for it was known that temple prostitutes died young. The account in Genesis tells of the sin of Sodom, which was so prevalent that not even ten "just" (healthy) men could be found to prevent the city's wasting.

If, as modern science seems to conclude, the A.I.D.S. transmission is in sexual exchange of body fluids, the possibility exists that the fluids may be more than just a carrier, the fluid's DNA may incubate (provide DNA) the retrovirus. In terms of the test of evolution (justification), the female system would seem to be more *justified* than the male because asexual reproduction (cell division) has been around much longer that sexual. The male sex is much more recent in its "justification"; thus, the symbioses brought forward in asexual systems (female cells) may be better accommodated and the female system may be more secured (justified) than the male. If this is true, the male system may be more easily infected by retroviral disease, and for that reason may more likely have had the disease develop in it than in the female system.

Common moral sense tells that it is contrary to right reason and to habits of good health to expose oneself and others to greater risks of disease predisposition by engaging in insulting behavior. It makes good sense to inform oneself in good biology for by so doing one promotes good health of mind and body, which is an exercise of good religion even as it is of civility. A good civil, theological premise to live by is the maxim that "good biology is good religion".

Toward 20-20 Vision

Perspective is distorted when vision is defective. When self-consciousness is too one-directionally focused, for example, in a "God-me" fixation, one's perspective tends to be distracted from other equally important perspectives of "God-other". Perhaps the most notable, and historically the most telling consequence of this defect is narcissism, a predominantly male proclivity of exaggerated self-estimation. An anti-social outcome of it is sexism, an exaggerated cult of male glorification whose other face is the put-down of females.

The theory proposed here is that the hyper-sexist, mono-directional focus of patriarchal theology has radically infected humankind with a diseased personal/social vision, and has habituated

Judeo-Christian-Islamic religions in a bias that distorts consciousness in ways that are ultimately manifest in violence of varied kinds.

Healthy vision is restored with a healthy bi-focal function, which originates genetically in split cell (bi-sexual) reproduction—nature's (God's) paradigmatic plan for balanced consciousness and personality. It takes two eyes for vision to be triangular and it takes the combined good sense of male/female sex for consciousness to be triangular. Isn't this religious fact-of-life really the root consciousness of vision into Trinitarian Community?

COUNTRY BOY

Child, chasing in tallgrass prairies,
Woods and forests, wading
The purling brooks, has little inkling
How much these are imprinted in him.

Native Americans have long known
And cultivated this truth in religious rituals.
Profound evolution, probably more than ever
We can know, is impressed and expressed
In the flip aphorism:
"You can take the boy out of the country,
But you can't take the country out of the boy."

The streams, the flora and fauna flow and flourish
In the human body. The tried and ancient
Use of Earth structures is engrained in body symmetries;
The psyches of Earth and body together
Move in natural harmony.

SECTION III

ON PURPOSE

PURPOSE speaks to the reason of things, to the intelligence behind the authenticity and sustainability of cause-and-effect relationships. Purpose is the glue of sense and sensitivity that connects the faculties of reason (intelligence) and faith (intuition). Knowledge of effects has everything to do with right and wrong, with good and evil, with making choices on a daily basis that have beneficial (sustainable) effects.

Purpose is rationally derived. There can be no sustained commitment to purpose (love) except commitment is based in faith and reason. Love can no more be separated from faith and reason than conscience can be separated from communication and consciousness. Love is the fruit of the marriage of Faith and Reason, even as conscience is the fruit of consciousness and conscience. Love, like consciousness, is the determination of *purpose*, the rational fidelity of cosmic intension and self-aware intention. In the end, only love's purpose endures.

Love's purpose is born of Faith and Reason. St. Anselm's insight "faith seeking intelligence" has the bride pursuing the groom. The authentic pursuit of love is mutual, purposeful, and so is the relationship of Religion and Science. The intuitive wisdom of consciousness (Fides) pursues new knowledge (Ratio) even as new knowledge must accommodate intuitional wisdom.

Integrity's grounding is faith-consciousness, the secure awareness of personal identity fitted essentially in the seamless whole of cosmic rationality. Reason seeks understanding of fidelity even as fidelity is secured in the understanding of reason. Personal and common well-being are either advanced or victimized by the choices we make and the actions we take; thus, "purpose", intention, the reason behind our actions has everything to do with morality, with ethical conduct. Faith and reason are wasted when actions are taken without concern for their impact (effects) on personal and social relationships. The morality of purpose, living with sensitivity for cause-and-effect relationships, demands vigilant effort to keep personal and social conscience informed—to keep faith and reason in proactive communication with each other.

A Purpose to Every Season

NOW is the acceptable time. NOW is the only time humans have by which to secure the present and thereby safeguard the future. There is the ever-present danger that The Now is lost in preoccupation with the past and future. How *Now* is used determines human and web-life destinies.

Common well-being corresponds with creation's proven ways of "holy purpose" (*sacra mens*). Human living like all living is subject to annual cycles of continuity, by which life originates, thrives, fruits and dies. It is to facilitate mindfulness of life's circumstances and to inform participation in them that life is *religiously* celebrated and respected, a worship of divinity that is primal.

Spring Equinox: A Time to Plant.

Spring is a time to plant. Consider the mystery of life packaged in a seed. The values we live by are seeds, for they are exemplified in our daily living and root in other people's lives. Enter into this mystery with full knowledge that good fruit is produced when human conduct conforms to the intended order of creation. Let us prepare ourselves now for planting that is faith bearing, for God is Provident when the order of nature is not abused.

The hope for fruitful planting is in the soil's ability to support the growth of life in seeds. Females possess an intuitive sensitivity to this mystery for it is in their own bodies that human life originates and finds growth. It is with informed purpose that men and women should act in preparing soil to achieve the successful growth of new life. Soil is the substance of life. Soil sustains human life for its duration and is enriched in the return of its materials with death. Soil is sacred because it is the substance of life.

Before there was soil, there was water and sun. Original life formed from water. In its proliferation life gave origin to atmosphere and soil. Soil is therefore both the creation of life and the substance of life. Soil is life's residue. Life presupposes water even as all Sacrament presupposes baptism, for in water's life "holy purpose"

175

originates and is accomplished. The substances of soil (earth-below) and atmosphere (heaven-above) are the substances of life, whose effective working is the sign of God's Providence. Conservancy of life is in the Trinitarian circumstance of life. Abuse of life's substance is sin for it wastes life. The Sign of God's Presence is in Trinity by which life originates and is gifted. The gifting of human personality in Family is also Trinity. This is the great mystery of Being, which is in God's Person and our own.

NOW is the acceptable time to reflect on these mysteries. NOW is the time to plant the seed. Consider the mystery of life packaged in living seeds. Enter into this mystery with sure knowledge that faith in God's Providence is not misplaced when living conforms to supporting the work of creation. Planting, which is faithfully practiced, that is, done with conscience (*cum scientia*) bears the fruit of sure hope for a better life.

Summer Solstice: A Time to Grow.
Summer is a time to grow. The meadows, fields and forests are bursting in green. Insects and birds are busy with growing offspring. Humans can plant, but only God can give increase. Without discrimination the sun shines and rain waters earth. The working of water in life is miraculously powered by the sun's energy. Life presupposes water as Sacrament presupposes baptism. Open your eyes to the Providence of God in the balanced working of nature. Discover the mystery of water and hallow it for the sacred gift that it is. Praise God as earth drinks the refreshing rain.

Growth is in two ways, from within (by intussusception's growth) and from without (by accumulation's growth). Growth of living things is from within, that is, within living cells whose every function is specifically orchestrated by the genetic *computer chip* within them, namely, the DNA molecule.

The production of materials in cells is from ingredients of earth and atmosphere, which are essentially the same in all life. The genetic code-materials of diverse life forms are themselves remarkably the same, whether in an oak tree or in humans. The structuring of cell materials occurs at the level of atoms and molecules, where electrons of different materials are exchanged and shared according to laws of physics and chemistry, and as programmed in the DNA code. New materials with totally new characteristics are brought into existence by electron sharing and exchange. The genetic structure of life is at the same time the preserver of life's unity and continuity, and is the source of life's great diversity.

The working of nature appears so effortless when conditions for it are right. Humans can do no better than to come to greater understandings of the conditions of the working of life itself. Such knowledge makes conscientious living more possible. When conduct is informed, humans can intentionally and effectively act to make the sacramental working of life more successful. That is really what farming is all about. Through informed conduct people come to a greater sense of the holiness of soil, air and water, and they come to promotion of life by conserving its sacred materials. Mindful cultivation of life means to enhance it, not diminish it. Above all, cultivation means not to destroy life. The obligation of forming a correct conscience in this matter is especially heavy on this generation.

Fall Equinox: A Time to Reap.

Fall is a time to reap. Fields are heavy with grain, the gifting of water and light. Grain becomes our daily bread, which, though not by it alone do we live, is nevertheless necessary for life. Waste of land and waste of food are desecrations of Sacrament for they waste life. All crave the bread of life, the Eucharist of life's common union, for participation in divine life is made possible by it. Eucharist presupposes baptism as grain presupposes water; the former is not possible except for the latter. Discover that all gifting of life is in the sacramental working of life working in the materials of earth and atmosphere, whence the garden is made fruitful and divine life becomes the experience of humans.

Fields heavy with ripe grain are the good fruit of good seed. The seeds of grain are both food for other life and genetic packages, which insure the life of its own kind. Seeds represent the unbroken continuity of present life with original life. Strictly speaking, life cannot be said to begin or to end, but to continue. Living seeds possess and pass on life's continuity by growing and by being consumed. Individuals appear on the scene, live their lives, and pass on. Individuality originates when a new life cycle is set in motion.

That which an individual brings to life is what is possessed in future life. Humans uniquely have it within their power to enrich or diminish life according to the way they use their intelligence in living. The values by which people live cause either good fruit or bitter fruit to be reaped. Bitter fruit grows from self-serving and from uninformed living. Out of knowledge of the truth comes the ability to live fruitfully. And fruit is even better born when personal conduct is informed, for committed conduct is rooted in knowledge. Informed commitment facilitates the sacramental working of life and enriches all life by causing the fruit of divine gifting to set.

By its fruit the tree is known; the authentic example of Christian living is shown in love for one another. Love is fulfilled in service. The call of service is made to every person. But more, service is the trademark of the Christian. It is the call of universal priesthood.

It is fitting for each of us to enter into this divine mystery so as to become a living Sacrament to others by our serving. In service is true, human fulfillment. A life of service is followed with a dignified and joyful death, free of remorse, for the gifting of such a life is unto untold generations. Death is no grim reaper to the Christian but a necessary condition for new life and life's renewal. Unless the seed die, it remains alone.

Winter Solstice: A Time to Rest.

Winter is a time to prepare and reflect. Living seeds must be put to rest to survive the harshness of winter. It is an awesome time and not without some worry that some catastrophe might befall the link of life now motionless in the dormant seed. It is well for us to understand the delicacy of this time and of life itself, and to better appreciate the circumstances of life necessary to unlock the seed's hidden powers. Human conduct like the growth of a plant is from the germs of value possessed in the human person. Winter is a time to improve the values by which we live. Enter thoughtfully into this mystery.

Sabbaths and sabbatical leaves are customs of ancient origin whose purpose is to refresh and restore. Today, people take "vacations'. The biblical account of creation has the Creator resting on the seventh day. Nature too has its time of rest.

The chill of winter causes living things to enter a cycle of rest, during which things happen in preparation for springtime when a new cycle of growth starts all over again.

For each living creature there comes a time to die. Each individual, having lived his/her life, must return to earth the materials borrowed where they are restructured and reused. The cycles of life are necessary because the energy of matter runs down and degrades. Renewal by recycling restores energies and reuse in nature. In the genetic character of each living thing the successes of past life and the hope of future life are neatly packaged. Life is in constant quest of bettering itself for acquisitions of new dimensions, but always using the same materials over and over again.

Life is the ladder between heaven and earth. Already in the Old Testament, Jacob wrestled with this mystery. As he was traveling to a far country in search for a wife he had a dream. He saw in his dream a ladder standing on earth and reaching up to heaven, and he saw messenger spirits freely coming and going on the ladder. Modern science has come to a new perception of Jacob's ladder. Science has discovered that the ladder is a spiral staircase. The DNA molecule, present in every cell of every living creature, is the spiral staircase, which programs life in its cyclical living. It is the most remarkable *computer chip* known because it advances codes of psychological/physical structuring that enables a body for its lifetime and for lifetimes into the future. Purposeful action is codified in the working of matter; conscious motivation identifies with matter's working, the Mystery of Sacrament.

Now is a time to prepare, a time to rest and read. It is a time to enter into the mysteries of life in thoughtful study, for with greater understanding of the working of nature we become better equipped to conform our personal living to the sacramental working of nature. Heaven and earth are equally home of mind. Through a heightened consciousness we are made to participate in divine life, for the working of Sacrament is the working of God in creation.

Wisdom's Word/Work

Unwilling visionary, I am yet passionate in advocacy for intuitional wisdom and fearful of where my passion might take me. My advocacy for cosmic wisdom is ambiguous, for part of me wants to preach from the housetops, while another part of me fears public exposure. Wisdom's awareness is an agony; it is a personal dying on behalf of common resurrection—as is witnessed through the ages. Aware of my ambiguous mind, I ask myself, *"What would you say if you were asked to share your life understandings with others?"* I would want to share with others what has made me whole. But how does one condense a lifetime of cumulative learning and experience in a coherently comprehensible way without boring the listeners? I might say something like this:

"Dear long-suffering people! Please be tolerant of me for I suffer as much being before you as you do being before me. In business experience, the audience I mostly talked to, has been farmers and agribusiness people interested in learning about corn grain life and how to preserve food (biological) values in corn seeds when stored—for seed vitality, its living qualifications, define the value-state of corn grain.

And, if we can take Isaiah's words literally, "All flesh is grass", it is altogether fitting to investigate the parallels of values associated with grain vitality and human vitality. Isaiah's "flesh" and ours is corn—grain—*grass*. The grain seed of grass is food and Eucharist. The community of people, the spiritual/material viability of people, the People Church, is essentially defined by valuated life's communal vitality—a complexity of cosmic homogeneity. God's house, grain's house, common ecology, is the Eucharistic place of God's People, the Church of Vatican II. Today, this house has for many of God's people lost its vitalizing power; many are alienated; communion service and communal life, spiritually and materially, are denied to many children in the house.

The cumulus of genetic study, especially, findings of the human genome project, is an encyclopedic documentation of the relatedness of all vitality, and of the common evolution and genetic constitution of all Earthlife. Human genes are made of the very same stuff as the genes of slugs and butterflies. And what has been a large come down from the expectation of conventional wisdom, humans do not have that many more genes than they do—something like only twice as many. And the genes of slugs and butterflies call on them to do many of the same things that ours call upon us to do. Wisdom is in the genes, sequenced over time in the crucible of chemical/biological trial-and-error; in a word, vital consciousness, wisdom, is the word/work of trial-and-error, of *experienced memory* and reflex/reflective rationality over the cosmic course of cosmic history.

At the common root of evolutionary process and progress, one gets a sense that the first, common, original channel of all evolutionary process is "communication", whose many functions perform in ways that correspond with the subtleties of the subjects-in-process. Thus, *"word" working* isn't merely about God "speak", creation working, it is also about God as *Word and Work* in the order of natural sacrament. The nature (work) of communication changes as the subtleties of subjects change. The more complex the subjects, the more subtle and complex are the ways and workings of communication. Word/working is sign/grace. Awareness of oneself as *sign/grace* is a most subtle and highly rational accomplishment of complex, *natural* communication.

In his book, *At Home in the Cosmos*, David S. Toolan gives us a feel for the dialogic process involving the evolutionary unfolding of Hebrew/Christian consciousness through history. He points out that Hebrew theology through history evolved through the dialog of two distinctive strands of thinking at work on public consciousness. Orthodox Jewish consciousness (Zionism) is political as well as theological; state and religion evolved from the unity of Hebrew two-strand rationality. The language and consciousness of the two strands are however, quite different from each other. The two differing voices are notably present already in the two accounts of creation in Genesis.

The first strand is the "Priestly", which tends to be backward looking (retrogressive). From the *Priestly* perspective, life in the

Garden of Eden immediately after creation was pristine, but after the First Parents committed the first (original) sin, grief and havoc came to all life in the Garden. All living relationships thereafter have been on a downgrade—until the death of Jesus on the cross, by which God's grace liberates humankind from the bondage of the Eden-sin, and humans are enabled to defeat their disgrace and expulsion from the garden. The second strand is the "Jahwist", which sees life as God's breath on earth, females on a par with males in God's mind, and all humans in a constant struggle to survive natural challenges. It looks to the future, confident that God is with humankind in its struggle, and that God suffers along with humans/nature.

The *Priestly* mind is "prescriptive", it is rule inclined. It formulates laws, rituals and commandments, by which human conduct is rigorously prescribed, whether in everyday matters such as cleansing, food preparation, in all aspects of living relationships and especially, in acts of worship and atonement. The *Priestly* strand is patriarchal and controlling. The man, Adam, is placed over all things, including wife and children, and presumes for himself divine appointment to dominate. The *Jahwist* strand is more sensitive, it is more feminine in its characterization of God's benevolence; it is more sympathetic to human weaknesses and is unconditional in its love; it is compassionate for the pain of people and not condemnatory of them because of human weaknesses.

As the *Priestly* strand evolved it became ever more complex and repressive in its prescriptions, whether in social or religious matters. Because of its rooted obsession in the past, it is more *regressive*. The *Jahwist* was not legalistic and exacting, it rather sought by divine example to lead people to right ways. Through history, the *Jahwist* vision has been future-looking (progressive).

The dialog of these polarities, the regressive (conservative) and the progressive (liberal), has instinctively and intuitively been at work in personal/social rationality. In religion and politics, these strands are present today, and probably will be for all time. On balance, both strands contribute to the rational weave of common well-being. The dialogic reconciliation of communication is the rationality of new consciousness. The *Priestly*, the prescriptive, and the *Jahwist*, the intuitive, together serve the harmonic function of conscience-formation and moral judgment.

The *prescriptive* exercises control over behavior that is socially harmful; the *intuitive* serves to affirm the authenticity of individual insight in service of altruistic relationships—a socially inherent and coherent purpose. Personal conscience, also, like societal dialog, functions between the pull of the prescriptive (what not to do) and the intuitive push (what to do). Harsh, prescriptive authoritarianism, when it lacks evenhandedness and acts excessively, repressively, can be an agency of distrust. Intuitional wisdom challenges the logic of excess and, thereby, works to affirm trust that is secured by reasonable restraints—the proper role of the *Priestly*. The *Priestly* might be characterized as male, and the *Jahwist* as female. Purposeful mutuality reconciles them.

The cultured excess of male-electionsim over femaleness is a strand of *Priestly* affirmation that manifests itself harmfully in virtually all aspects of civil institution—to the extent that violence and even the psychological malformation of the young become habituated. Male-dominated religion (society) is a self-appointment of male egoism. In the domination mind of males, women and children are treated as property. Male children are rationalized to be in the image of the Godhead, whereas, female children are rationalized to be defective males by reason of women's default. Even the male-conceived godhead (trinity) is cast in male-electionism, in the father/son "procession", whose mediated consciousness (male) is the Holy Spirit. The discordant mind of narcissistic self-preference is exposed for what it is, possessive/obsessive male-idolatry.

The *Priestly* culture has evolved a tight hierarchical, authoritarian lineage, which, it believes, ministers the spiritual virtues and the social mechanisms (laws) necessary for personal salvation, well-being. As in the Hebrew culture, so many now believe also about Catholic Christian culture, the dominion of *Priestly* orthodoxy has become excessively controlling and suffocating.

When the suppressive hand of prescriptive control becomes too heavy, the intuitional voice of the *Jahwist* strand, out of compulsion for survival, is provoked to stand up against it. The prophet, who sees through the idolatry of harmful fixations, calls for a release from the oppression and for the healing of the people (Sabbath) and the land (Jubilee). God's voice is identified in the promise of liberation, justice and hope. The dialogic crucible of people experiencing hurtful prescriptions extracts new and better insights in the service of altruistic relationships. Entrenched *Priestly* orthodoxy resists the prophet because the prophet challenges its credibility and accustomed authority. The prophet is revolutionary, evolutionary, and sometimes even anarchistic, if entrenched domination does not relent.

The history of protests within the Roman Catholic Church, say from 1400 until now, can be characterized as a war between the minds of the "Catholic Apostolic", *Priestly,* and the "Catholic Prophetic", *Jahwistic*. As long as the prevailing *Priestly* theology was rooted in a publicly accepted worldview, it prevailed. The accepted worldview assumed a cosmology that was centered on earth and mankind, and, a God, above and outside, the earth-centered realm. The prescriptive order of *Priestly* consciousness evolved from the same presumption, and priests, by divine appointment, mediated between the people and God "above". However, that worldview has totally collapsed, as well as has belief that God is alienated from the people except through the linkage of the hierarchical priesthood. Understandably, the mind of *Priestly* control in Church is presently unsettled and undergoing a sea change.

The Prophetic Protestant Reformation was driven by two principal factors, gross corruptions within Apostolic Catholicism, especially, its obsessions in power, wealth and kingdom building, and by growing public awareness that neither humankind nor earth was the center of the cosmos. Among early thinkers who had worked out a "new cosmology" were Copernicus, Bruno and Galileo. The laity (reason) challenged the clergy (faith).

In its counter-attack against the Protestant Reformation, Apostolic Catholicism attacked the science and the persons of the *new cosmology*, for it knew well that if the new cosmology gained public acceptance, and the established worldview lost public acceptance, then the rationale for and the whole body of its theology would be at serious risk of losing acceptance—and, the Protestant Reformation would be vindicated. So, Courts of Inquisition were established in Spain and in Rome, and people of non-orthodox beliefs were brought before them and were tortured so as to make them recant and return to orthodoxy. Not to recant would result in condemnation and death, even burning at the stake, as was Giordano Bruno's fate.

Roman Catholicism reacted with a further tactic of defining even more strictly the tenets of Apostolic Faith, thus, at the same time, redefining and re-hallowing its ancient cosmology and the tighter sanction of its male, hierarchical priesthood, including papal infallibility in the matters of beliefs and morals. This tightening of discipline occurred in the Councils of Trent and the First Vatican.

In biological science, there is the saying: "Ontogeny recapitulates phylogeny". The individual genetic code draws upon the genetic pool of all humankind. The vast ocean of human insight and wisdom resides in the whole gene pool; the insight and wisdom of the individual is only a "drop in the bucket" by comparison. Thus, wisdom and faith reside more completely in the whole "pool of humankind" than in the "drop", the individual. The logical conclusion of this understanding is that faith is better placed in the "sensus fidelium", in informed public consensus, than in an individual authoritarian figure.

Priestly infallibilism is counter-intuitional, and the absolute submission of faith to its reactionary harshness is unreasonable on its face. Consciousness tells us that the "prescriptive voice" is external to one's person (ad extra), while the "intuitive voice" is personally internal, within (ad intra). Informed personal conscience is a higher court of law than institutional prescription. The voice of informed conscience is certain, whereas, the motive of the prescriptive voice is less certain. The voice *ad intra* is intuitive, *Jahwist*, whereas, the voice *ad extra* is institutional, *Priestly*. The *Priestly* Church of Vatican I (Tridentine) is prescriptive, Petrine, whereas, the *Jahwist* Church of Vatican II is intuitional, Johaninne. (See Michael H. Crosby's "Do You Love Me: Jesus Questions the Church", Chapter 12, 2000, Orbis Books, Maryknoll, NY.)

Today the *Priestly Apostolic*, Hierarchical Church and the *Jahwist Prophetic*, Discipleship Church are fighting it out. While there is yet a lot of frustration and hurt being suffered by the faithful because of internal polarization, intuition is likely to predominate because the old underpinnings of the hard theology of the *Priestly* is without the cosmological base it once had. In their religiously instituted self-arrogation over females, males expose themselves, quintessentially, to be distrustful. In its institutional incorporation of patriarchy, institutional Church claims divine sanction for male super-arrogation.

Patriarchy originates and explicates the *Priestly* strand in Judaic/Christian tradition. In the *Jahwist,* wisdom-strand, God, not male super-arrogation, is seen as unconditional lover instead of a harsh exacter of hard law. *Priestly* male love is conditional, legalistic, and contractual, whereas, God's love, like a mother's, is non-contractual, covenantal, and unconditional. The logic of the *Priestly* strand is the logic of a lawyer, whereas, the logic of the *Jahwist* is the logic of a faithful lover. The mind that is covenantal presupposes trustworthiness, whereas, the contractual mind of the lawyer is instinctively distrustful.

Some people are of a mind that priests and bishops should be required (at least allowed) to marry, for male accommodation to female wisdom is good for the soul, good for common well-being. The man who is seriously committed to be honest as a husband and parent is a person not likely to be duplicitous, for duplicity with them destroys their trust and frustrates family. A person cannot be a good parent and be duplicitous and distrustful at the same time. The celibate hierarchy seems able and willing, in the interest of institutional fidelity, to act duplicitously toward non-institutional people, especially, toward women. The ordination of women, and a married priesthood, might help heal the deep breaches of trust that now exist between lay (especially female) and clergy.

Bringing an end to abusive relationships is an individual responsibility, regardless of the professional position of any individual. Word/work is the process of wisdom, of dialogic exchange between the rationales of closed-ness and control, and, of openness and conscience. Violence between the orthodox and the non-orthodox can result only in more violence and more societal contention. Accommodation to intuitional rationality, to symbioses, can mitigate violence and misguided conduct. Symbiosis can work its good only when the *Priestly* and *Jahwist* minds reconcile in the common good.

But, the atavistic *Priestly* sentiment of self-arrogation seems to have become a crusty, layered accretion of prescriptive harshness in the hierarchical church. Like its Hebraic antecedent, Tridentine Catholicism means to ensconce permanently its fixations in the sediments of canons. However, the *Jahwist* Spirit of Vatican II means to peel away the rock hard layers. Regressive thinkers and progressive thinkers need to reconcile on common faith/reason grounds. "Whole making", salvation, is the work of the universal priesthood of service.

Everyone is born into the mission of the priesthood of service. Not at the extremes of hyper-charged polarities, but in the middle estate, are vitality and viability sustained. The mitigation of fractious polarities is *sacrament*, is symbiosis, the "sacred purpose" of the natural inherency of vital good will.

Authoritarian religion (apostolicism) confuses its role with the role of civil government. Jesus' witness requires more of disciples/apostles. The political bureaucracy of civil government is not normative for jurisdiction in Church because civil jurisdiction pertains to maintaining civil order. "Church" people are presumed to be civil, people of faith and good will. Apostles must be more than controlling hierarchs; they are called to be disciples who motivate others to loving relationships, firstly, by exemplifying love themselves.

Vatican I Catholicism seems to have the mind that *apostleship* (discipline) and *discipleship* (prophecy: inspiration/motivation in love) are exclusive charisms. Vatican II Catholicism seems more of a mind that they *(discipleship/apostleship)* can and should be inclusive, that is, *not* mutually exclusive. While the Petrine Church, like Peter, comes late to love-discipleship, the Johannine Church, after Jesus' example and mandate, challenges it, *"Do you love me?"* Apostle/disciple, all are called to service in God's word/work, the inseparable marriage of love/labor. What God has joined is not to be separated. Jesus tells us, "Go, live personally the word/work of catholic apostleship/discipleship in the bond of the Faith/Reason covenant."

Purpose

Intension & Intention. In their perennial dialogue, philosophy and theology become enmeshed in the matter of the "essence" of things, their "quiditas", that is, the *what-ness* of everything. Out of the gymnastics of psychic implication evolved the hide-and-seek process of coming to knowledge (scientia), which underlies and drives human self-reflection and intention.

The imbroglio of this process becomes sometimes so impenetrable as to lead to divisive fault-lines and unreconciled differences. The cleaving of consciousness along fault-lines gives rise to all manner of unanticipated and unintended consequences, for the reality is that we are individually creatures of our own prejudices of *intension and intention.*

Energetic intension/intention is the dynamic basis of physical/conscious reality. This observation provokes a question that is radically logical, if it is true. Is "purpose", whether or not reflective, a conscious complexity of intension/intention? And, is purpose inherent in the directionality of evolutionary consciousness? I would say yes; and even further, I would say that the question is of particular importance because it goes precisely to the subject matter of "cosmic rationality". Cosmic rationality pertains to the intelligence that we ascribe to the purposeful direction of evolutionary transformation.

While the intelligence of a child is at birth for the most part yet *potential*, that is, still infolded in his/her unexpressed genetic make-up, no reasonable person would deny that a child is *purposeful* in his/her exercise of rationality, even as an infant. Going back one step further, a reasonably informed person recognizes that the fertilized ovum, implanted in the womb, is driven by its programmed intension, whose *purpose* is specifically directed in the explication of its embedded genetic potentials. And to go back even further: the purposeful organization of the human body enables specialized cells functioning as specialized organs to purposely produce specialized cells that carry within them the purposeful potentials of the *essential human* and to put these cells in necessary communication/communion with other cells and together accomplish the purposeful bringing into existence a truly new, intentional human being.

This self-reflective exercise of reverting consciousness back to deeper experiential considerations of purposeful self-awareness is itself instructive of the cosmic reality that we are.

The tensional inherency of static electricity stores potential in particulate substance. Static charge in particulate substance is fundamentally purposeful in initiating and sustaining the directional success of interactions. Sustainable interaction is a *purposeful* basis for interactions, which increase particulate capacities for ever-expanding interactions. Sustainable interaction and increased capacity for new interactions are obviously purposeful in promoting evolutionary transformations. The (in)tensional inherency of first substance is the aboriginal, quantum-electric base of *intentional relationship—religion*. Except for inherent particulate tension there would be neither sustainability nor purpose. Purpose in cosmic continuity is self-evident, *axiomatic*.

In science it is axiomatic that gross manifestations of Big Bang phenomena appear more refined in later complexities that evolve in essentially continuous transformations. If the axiom is true, then we should be able to reflect upon ourselves and discover not only more about ourselves but also more about the cosmos itself and the inherent directionality that has been implicated in its energy/matter potentials from the beginning. In cosmic grounding, religion (theology) and science (philosophy—cumulatively associated knowledge) find resolutions to common questions that are customarily and opposingly viewed from self-interest perceptions and assumptions.

A further instructive axiom is that all cosmic manifestations explicated by evolution in new manifestations are already possessed potentially in the cosmos whence they come. In that we come from the energy/matter of the cosmos, the cosmos possesses the matter, the rationality, the intentionality, and the *purpose* that comes to be ours. All rational, *purposeful intelligence* is possessed in cosmic rationality, in potential, which participates in and expresses *divine purpose* as it is carefully parceled out in directionally evolving matter/energy.

Atomic/molecular intension is a quantum-electric, phenomenally qualifying relativity involving and evolved from complexities of electrical fields and particulate attraction/repulsion. As such, intension is an essential edification phenomenon of atomic/molecular assemblies and potentialities. Atomic/molecular intension is the relativity precursor of intentionality; philosophically, it may rightly be said that it is axiomatically conclusive that *intension* is a root phenomenon of cosmic consciousness *later expressed as intention*. The evolution of self-consciousness is part and parcel of cosmic rationality. It is a psychic phenomenon of reality, imprinted in quantum-electric patterning, whose potentialities are not wholly definable, knowable and predictable. Dare we say that divine hopefulness drives and is mercifully, lovingly embedded in the surprise rationality and potentiality of cosmic intelligence?

God is subtle even as the evolution of purpose and the purpose of evolution are subtle. Subtlety accommodates the *ad hoc* circumstances in which purpose becomes manifest. As in the unborn Child, cosmic origins possess, inherently and potentially, purpose and self-aware rationality. In the driven consciousness of purposeful rationality all life and the cosmos itself are driven to achieve fullness, "pleroma," in Chardin's word. Heaven, at-home-ness in divine presence, is the ultimate relativity of all existence, and is the eventual and ultimate achievement of self-reflective consciousness. From and in this reality we are born, and in it we thrive, and to it we must return. The lesson is that cosmic rationality implants within us the purposeful necessity of becoming conscionable creatures, that is, creatures informed in our origins and intentionally conformed to the harmonic resonance of their wisdom. The motive of essential conscience is Love, the ultimate call of perfection that speaks from and within Nature, whose mystery continues to deepen even as its infinite history continues to unfold.

Subsidiarity and Hierarchy. As phenomena of naturally evolved order, subsidiarity and hierarchy are correlatives in ongoing evolution and in the socializing of highly advanced life. Subsidiarity links the agents of continuity, whereby established agencies join and transform to become new agencies and establishes a continuum of order wherein all agencies are *hierarchically* related, that is, new agencies originate from and depend upon prior agencies. Nature doesn't *reinvent the wheel*, which is to say that evolved mechanisms of change become embedded and workable in newly developed agencies. Subsequent agencies are *hierarchically*, wholly dependent on prior agencies.

As new and sustainable mechanisms are perfected, ongoing transformations depend on the cosmically proven process of subsidiarity, on the new bases of change embedded in prior transformations. Circumstances still arise that challenge biological success and require resolution. The *ad hoc* resolution of challenging circumstances by the entities involved in them and by accessible means at hand is the most workable way of dealing with them. Churches and political structures subscribe to the principle of subsidiarity in their social functioning, however, entrenched hierarchy, that is, bureaucracies of control that are patterned after monarchical (top-down control) absolutism, do not easily let local churches deal with local circumstances involving them.

Presently within the Roman Catholic Church a contest between Roman Catholic centrism and local subsidiarity continues. It is a contest between the philosophies of Tridentine Catholicism (the Councils of Trent and Vatican I) and Vatican II Catholicism. Tridentine Catholicism insists on micro managing the Church through hierarchical bureaucracies at all levels and in all matters, whereas, Vatican II Church (The People Church) urges greater local latitude and lay involvement in handling local Church matters. The contest is between the Curia, the central control bureaucracy of the Church, and the local bishops who are appointed and ordained by the Church to exercise local authority in Church matters.

Perhaps it states it somewhat simplistically, yet not inaccurately, to say that the centrist/regionalist conflict is an ancient and perennial one, one that existed also in the early Church, namely, under the opposite personalities of St. Peter (the Church of centrism, power and domination) and of St. John, the beloved of Jesus—the Church universal, loving and tolerant.

In his theology of the "mystical body", St. Paul compellingly makes the case for functional diversity in the Church as in the human body. He elucidates the commonsense logic of natural subsidiarity and cosmic rationality. There is one body, one Church, and one Christ. In this one body there are many members, cells, and organs that contribute necessary and different services in the function of the whole body. The eyes cannot discount the work of the ears. The arms cannot discount the work of the legs. The head cannot discount the work of the heart. Neural cells cannot discount the work of muscle cells. Kidneys cannot discount the work of the stomach. Sperm cannot discount the work of the ovum. Nuclear DNA cannot discount the DNA of plastids and mitochondria. Reflective rationality cannot discount reflex rationality. Males cannot rightly hyper-rationalize their agency and repress essential female agency—in Church. In God we are one people, one Church. Before God we are, neither Greek, nor Jew, nor Islam, nor outcast, nor preferred, nor male, nor female. One and all we are one body, one Christ; each person is individually called to the mutuality of love's common service.

The principle of subsidiarity subscribes to naturally proven diversification, i.e., it allows for the diversities of regions and the authority within them to deal locally with the challenges of local circumstances, whereas, political centrism abhors diversity for "efficiency" reasons. Psychologically and practically, the narrow prescriptions of centrist monarchism are unnatural and unworkable, as is testified globally by the common distaste for universal sameness. Centrism and regionalism both have certain merits, and it is up to people globally, locally, to discover and authenticate the complementarity of contributions they offer.

DNA, life's blueprint, codifies all vital texts of subsidiarity and "hierarchy". In its original meaning, "hierarchy" implies "nothing other than that all things have their origin or principle (*arche*) in the domain of the sacred (*hiero*)." [Seyyed Hossein Nasr, "Islamic Cosmology: Basic Tenets and Implications, Yesterday and Today", Science and Religion in Search of Cosmic Purpose, edited by John F. Haught, 2000, Georgetown University Press, Washington, D.C., pp. 42-57; John F. Haught, "Information and Cosmic Purpose", ibidem, pg. 106.] The sacred purpose-*principle*, *sacra mens*, states the "sacrament" order of all creation, of transformational nature and cosmic evolution. All creation, no matter how deep the linearity and complexity of causal agencies, ultimately reverts in its origins to priority, to *divinity*. The divine source of all creation, of each and every hierarchical *subsidium*, makes all of creation *sacred* because every transformation, every evolved complexity, ultimately links back to and depends from/upon divinity—in *essentially* the same way. All soul/substance originates in/from One God, the Divine Agency operative in cosmic transformation.

Institutional religions, structurally premised on presumptions of self-rationalized divine right, have established "hierarchies" of domination and control over people, and in their political function disregard the egalitarian rationality/sanctity of all human beings. Change away from prejudicial hierarchical presumptions will not happen so long as they remain unchallenged and are left in place to assert prejudicial control.

Physical/Biological Purpose

Creation and life, like thought and consciousness, are works in progress. The study of physical science is a study of quantum relativity, the underlying "text" of biology. The study of evolutionary science is study of "contextual" transformation, physical and biological. Physical "intension" is an agent of purpose no less than is conscious (biological) "intention". Physical intension asserts its necessity in cause and in effect. Nature's harsh excess of energy, when in conflict with intentional vitality, is unremitting and forces accommodations on biological intention. Physical intension is an evolutionary cause-and-effect dynamic, an aspect of Darwin's survival-of-the-fittest evolutionary process. Accommodation to physical nature by physical agencies within nature causes change, evolution; and, genetic openness (<u>textual</u> mutation) is biology's mechanism for accommodating to <u>contextual</u> necessity.

Taken in context of each other, the science of quantum relativity and the science of evolution weave together insights of energy/matter implications in cosmic purpose and reveal the unitary origin of everything from the essential mutuality of energy and matter. As the explicit manifestation of implicated cosmic potential, inherent purpose, both in the physical and in the biological, tells of the essential continuity of intensional/intentional cosmic purpose. Cosmic physical order makes sense in its purposeful participation in biological evolution. The intensional order of physical nature underlies biology's web of intentional complexity; not merely "underlies" but <u>constitutes</u> biological intention. The imagination is in fact amazing "starlight".

Francisco J. Ayala ["Darwin and the Teleology of Nature", <u>Science and Religion in Search of Purpose</u>, edited by John F. Haught, 2000, Georgetown University Press, Washington, D.C., pg. 39] makes the unqualified statement that "biology cannot be reduced to the physical sciences". This statement can be appreciated as a defensive one against the sometimes shrill and simplistic reductionism by some scientists and schools of secular humanism. Surely Ayala knows that all living things and the science about them (biology) are causally dependent upon physical phenomena, chemical and energetic, and that biology has evolved from the subtle complexities of the physical milieu of energy-tensioned chemistry.

On its face, Ayala's statement seems intended to reinforce the traditionally held dichotomy between spirit and matter, soul and body, and, by implication, to argue for the extra-natural intervention of inspiration and revelation outside of explainable, causally related, physical interdependency. Arguably, in the light of quantum relativity and evolutionary dynamics, the unfolding of consciousness within the transformational universe doesn't require "extra-natural intervention" to account for it. Some might even say that the presumption of such intervention is wholly gratuitous and counter-intuitive.

"Science" (scientia) in its root sense means *knowledge*. Can biological knowledge be understood in terms of highly complex physical knowledge? Perhaps the claim for connection of physical causality in biology cannot be categorically claimed, but that does not mean that biological processes do not have their origin in "physical purpose", especially in light of absolute biological dependency on physical processes. In the least, the answer to the question is open as to the ways that the physical and the biological are dependent, and how the knowledge of each is distinguishable but also identifiable. Consideration must be given ultimately to the chemical/energetic constitution and relationships of biological systems, their ultimate sourcing in and interplay within physical "mechanisms".

Is it credible to postulate that inspirational, revelatory knowledge, phenomena of human consciousness/reason, are "purposefully" sourced outside of cosmic "energy", outside the *natural* train of physical/psychical phenomena? While Ayala seems to say "yes", Teilhard de Chardin would seem to suggest "no". In the above article (pg 36), Ayala makes an off-hand reference to Chardin, in which he facilely dismisses Chardin by associating him with the philosophies of Berg (author of "Nomogenesis or Evolution Determined by Law") and Osborn ("Aristogenesis, the Creative Principle in the Origin of Species"). Chardin's "evolutionary philosophy", according to Ayala, is erroneous because it holds that "evolutionary change necessarily proceeds along determined paths". Ayala rejects that biology is "physically" determined ("reducible") and seems to imply that divine intervention is the only other alternative by which biology is explainable. If *extra-natural purpose* (intelligence) is responsible for purpose in biological evolution, can it be explained by a way other than divine intervention, Divine Intelligence? Ayala seems to imply that "evolutionary change necessarily proceeds along [divinely] determined paths" that "cannot be reduced to physical sciences".

Islamic cosmology includes an inflationary notion of the divine in continuous creation. This sense of God has resonance with the creation theology of Judaism and Christianity wherein God "breathed" life into Adam. In Islamic thought God continuously "breaths in" the universe, and by which it is returned continuously to its "Divine Principle"; the outflow of breath expands and re-creates the universe in unending renewal. According to Persian Sufi philosophy "the universe is being destroyed and re-created every moment by God. Like the two moments of breathing, there is constant expansion (*bast*) and contraction (*qabd*) of the universe. At every moment everything returns to the Divine Principle and is then 'returned' and re-manifested, because if left to themselves, contingent beings would immediately collapse into nothingness...For the Sufis, *'adam'* refers to the celestial archetypes upon which God 'breathed' the Breath of Compassion...by which these archetypes become existentiated in outward forms".

Time is seen more as a quality of the creative universe. Even as cosmic regeneration is cyclical, so is time. Change in time registers in the amplification and attenuation of the Divine Principle. In time's cycle the process of universal renewal is a process of perpetual return. So it is, ships return to their original ports, as do trains to their stations.

Cyclical regeneration occurs at the level of atomic/molecular function and structure; even the human body is periodically and totally re-substantiated. Vital re-substantiation is a process of "resurrection" even as it accommodates to physical contingencies. Perhaps for most people over time, physical causality does not adequately account for biological teleology, even though personal experience tells that physical causality is totally involved. At the DNA level, the options of atomic/molecular openness enable genes to change (mutate) over time and accommodate to contingencies.

The capacity of physical substance to give life its vital potential is evident on its face, for that is what the digestion and assimilation of food is about, what breathing is about. At the atomic/molecular level, transubstantiation occurs. The dimension of the *spiritual*, the energetic, prevails in the physical as well as in the biological. The fact, of the presence of "spirituality" in physicality, cannot easily be dismissed by putting the name "reductionism" on it, rather, it needs to be understood in terms of the *essential relativity* of unity (homogeneity) that accounts for the constitution and the re-creation of the cosmic continuum.

Biology cannot be separated from its essential origin and sustainable basis in physical energy/matter. Nor does it make sense to rationalize mechanisms of biology as if they can logically and "really" be disconnected from the mechanisms of physics. Based on experiential and scientific evidence, it seems to make more sense to conceive of God, the Creator, as intimately involved in creation. Cannot God be conceived of as being *hands-on* in creation, in the sense that all creative agencies, physical and biological, are of common origin in divinity, and, therefore, all "holy" and part of the one and same essential, self-expressive continuity; and, to conceive of God as *hands-off* in the sense that creation remains open to include

196

infinitely contingent possibilities, physical and biological, and free to respond to contingencies in infinitely creative ways?

The artificially fragmented thought-processing engaged by humans is useful in breaking natural complexes down into bits and pieces for the purpose of coming to fuller understandings, but, in so doing people shouldn't come to the conclusion that piecemeal understandings of cosmic continuity describe a cosmos that is in fact fragmented in the way that it is imagined. People (all of us) do seize upon fragmented knowledge in selective and piecemeal ways that fit ideologically favored worldviews, whether for personal and/or institutional reasons. Notwithstanding our small-minded grasp of cosmic purpose, infinitely free-wheeling, it will continue undeterred, and it is in our common interest that we remain mindful of this, aware of the delicate balance that allows us place in the scheme of things— and that our personal claim on the physical substances of the cosmos is but an ictus in time. It is foolhardy to arrogate an importance to our selves that is not warranted in the Big Picture—better to know and honor Nature's greater Wisdom in serving the greater well-being.

If one doubts the totally controlling role of the physical world over the biological, stop and think: everything lives or dies in response to heat and cold; too much water, too little water. Our lives are regulated by day/night cycles, by seasonal cycles. Life depends on photosynthesis—sunlight—wave and particle. All the above are physical, energetic factors, and, not to forget, our bodies are physically reducible to the "dust of the Earth". All mass, living and nonliving, is energy-tensioned and capable of being reduced to energy.

If we live by our small understanding of the moment, we go through life without really being grounded. It is important to recognize the value of "purpose", the good sense of common purpose, for, such focus allows us to sort out the ephemeral nature of our changing awareness and to affix the bits and pieces of acquired knowledge to a grounded framework, thereby enabling us to find our place in the larger scheme of proven purpose.

Consciousness is a steamship named desire on a sea of infinite consciousness, a secured vessel in insecure waters. Self-conscious thought is a current of purpose tracking through chaotic terrains, a train straining to carry passengers and baggage on safe tracks to the destiny of desire. Steamship and locomotive are both destined to return to harbor and station of origin, and to swell the tides and trends of consciousness yet to come, serving purposes that only time can tell.

Network life is a maze of steamship lines and train tracks streaming crisscross within the cosmic "sea of infinite substance" (St. John of Damascus" definition of God), fulfilling intention's infinite desires and inflating imagination with creeds and ritual exercises at the harbors and stations of accomplishment.

Reason and faith steer with purpose the parallel and divergent courses of desire and destiny; one must secure the other, else destiny and desire are at cross-purposes.

The Rationality of Conscience

ACCOUNTABILITY FIRST HAPPENS IN PERSONAL LIVING BEFORE IT CAN IN INSTITUTIONAL. NEVERTHELESS, THE EVERYDAY ISSUES OF PERSONAL CONSCIENCE GET CONFUSED WITH THE PRESCRIPTIVE MINUTIAE OF MERCANTILE CORPORATISM. INSTITUTIONS REGULARLY LIE TO US, AND, NOTWITHSTANDING THEIR UNMISTAKEABLY CONSUMERIST INTENTIONS, WE MASSIVELY FALL FOR THEIR PROSTITUTIONAL PLOYS.

LEST I SEEM TO DISSEMBLE I MUST CONFESS UPFRONT THAT I TOO AM A CAPTIVE CONSUMERIST. WE AMERICANS HAVE BEEN BIRTHED ON THE RUNAWAY TRAIN OF CONSUMERISM AND WE ARE BLIND TO WHERE IT IS TAKING US. AS THEY SEEM TO OUR EYES TO DO, THE PARALLEL TRACKS CARRYING THIS MAD MACHINE CONVERGE AT A DISTANT HORIZON WHERE THE PASSENGER CARS WILL PILE UP IN A CALAMITOUS HEAP. IT'S ALREADY HAPPENING. THERE MAY YET BE TIME, HOWEVER, THAT, IF ENOUGH PASSENGERS WAKE UP TO THEIR DIRE SITUATION AND APPLY CORRECTIVE MEASURES, THIS WRECK-BOUND TRAIN MAY BE SLOWED DOWN TO A SUSTAINABLE SPEED THAT AVOIDS THE ULTIMATE WASTE OF EARTH RESOURCES AND THE GLOBAL COLLAPSE OF NETWORK LIFE.

Judging from the present direction, or misdirection if you prefer, of civilizations, it should be asked, "What is the future prospect for humankind?" If the global state of civilizations is a reliable indicator, the future for humankind is, indeed, bleak.

The history of human exploitation of Earth resources and network life seems to foretell more of the same and even more dread consequences for all life on Earth. Industrial pollution blankets and invades every aspect of Earth's environment, including all life forms and humans themselves. It seems accurate to say that the word "pristine" can no longer be applied to any region on Earth. The evidence of pollution is everywhere, and it shows up in new diseases that root in the aggravated trashing of natural defenses; consequences ranging from increased incidences of cancers to HIV/AIDS. A new threat to natural immunities, perhaps even more consequential than any previous threats, is the threat of mass disruptions of natural genetic codes by the human artifice of gene-splicing. And this threat is from none other than the corporate conglomerates of petrochemical companies. Perhaps more than any other agency, petrochemical conglomerates have fueled, and continue to fuel globally, the precipitous destruction of Earth life.

Even at this late date, however, not all hope is lost if humankind, the uncontested behemoth of havoc, can face up to its guilt and be honest in assessing the cumulative consequences of corporate misdirection. Finger pointing and recriminations are not purposeful at this late date. For example, religion should stop blaming science and science should end its dismissive attitude toward religion; the instincts for both religion (faith) and science (reason) are authentic and are premised in the experiential consciousness of cosmic rationality. Humankind needs, therefore, to evaluate both with respect to their mutual contributions to the present sorry state of human affairs, but also (!), to the collaborative role they can play in bringing civilizations to a higher rationality of consciousness—conscience. Conscience is a process of reason, a faculty of consciousness that controls personal judgments and actions based upon understanding and prioritizing actions according to their consequences.

Personal conscience is fatally flawed when it puts individual interests against the interests of others. Historically, theological emphasis in the Western world has tended to sensitize personal conscience disproportionately in God-me relationships and to the neglect of the other-me relationships. Cultural narcissism has been religiously promoted together with a dominion-mentality toward nature. Personal conscience, cultured in the presumption of dominion-right, is flawed with a mortal myopia toward its essential dependency on the sustainable necessities of Earth's network life. Commonsense rationality compels religion and science, faith and reason, to come to an accommodation with each other in personal and public consciousness, if for no other reason than for the love of self.

Where to begin with this accommodation? Perhaps at the beginning of the universe, with the Big Bang, for the ancient mythologies of various cultures yet prevail religiously in expectations of literal belief. These mythologies are no longer useful in their literal understanding, for the informed understandings of science now know better. The reason that the updating of creation mythologies is critically important is that the theologies that have evolved from them constrain habits of belief that are mistaken in their presumptions, and that, therefore, contribute to destructive oppositions of divided belief. One needs only reflect on the history of religious, ethnic and national conflicts now waged with greater virulence than ever before. Can it now be agreed that humankind is one in origin, one in brotherhood/sisterhood, and one in destiny? Can it not be agreed that all peoples, of all times, all individuals, are equally "chosen" by God, and that we cannot, personally and truthfully, as groups and as individuals, presume ourselves, our cultural history, to be more chosen than others? The mythologies of *chosenism* give rise to theologies of cultural, religious arrogance that drive nations in their presumptions of dominion right over Earth resources and other people. Witness the turbulent history of the Jewish people and the more recent genocidal havoc of colonial Christianity. Theologies need a new set of presumptions if they are to help humans extricate themselves from their long history of cultured fratricide.

Einstein has established for our edification the radical identity between mass and energy, between transformational matter and the qualifications of energy that give it its highly varied structure. The identified transformational character of embodied energy is called "quantum relativity". Quantum relativity characterizes the "essential continuity" of energy/matter transformations, whose origin is from cosmic decentralization that was caused by the Big Bang. *Essential continuity* continues even now in the codependent complexities of all energy/matter. The "rationality" of cosmic continuity in its present complexity, including Earth and all life on it, is the coded logic behind all laws of Nature. It is evolved rationality that now compels personal consciousness—personal conscience. And what does this tell theology about "divinity" possessed in all creation, about the sanctity and sacrament-nature of all creation? How is it that humans arrogate conclusions of God's presence more in some places and things than in others?

Cosmic rationality, the basis of conscious relativity, of conscience, equally obliges the conduct of individuals, and groups of individuals, for their lifetimes. Individual conscience can neither be corporately usurped nor personally abrogated. Each of us is individually responsible to be true to his/her personal rationality—conscience. Simply stated, Jesus' mandate "to love God and one another" is a statement of the communal obligation of personal conscience. In this mandate Jesus is revealed as the Cosmic Christ, for, the Christian "messianic" witness is the same as the mandate of cosmic rationality.

It is the cosmic and the Christian mandate that parents care for each other, and pass on to their children, the torch of love, of altruism, that motivates them in all their relationships. The rationality of conscience is *sustainability*; the logic of love is *altruism*—equal concern for other as for self. Love and altruism are the word/work of communal sustainability. The love parents bring to family relationships in raising one another to a more highly informed consciousness/conscience will do more to convert global civilizations away from corporate waste of life and resources than any amount of political posturing by institutions. If conscience prevails in nuclear and extended families, it will infect all bodies of civil society. Hope for the commitment of societal institutions to conscience, to accountability, lies in family commitment to conscience. Family is church. Church is community. Community is society. Societies are civilizations. Also, every complexity of humankind is essentially a "molecular" composite of "atomic" components—an energy/matter unity. The primary focus of theology begins with the nuclear family, even as the primary focus of quantum relativity begins with the atom. As the nucleus is the ground state of the atom, so femininity is the ground state of family, society. Fundamental social energy is a nuclear gravitation before it is societal.

Personally, socially, globally, who and what are we? Altogether we are many different things, different in talent, interests and motivation. But, one thing we are in common is a "quantum-electric molecule", cosmically informed. In every least particle of our body we are charged with electrical potential. Potential breaks out in consciousness, and consciousness breaks out in purposeful thought, action, artistry, etc., all of which result from the driven energy of conscious *intention*. The conscious intention of body potential (electrical *intension*) is resourced in every atom and combination of atoms composing the body.

The crossroad of conscience, the intentionally charged current of consciousness, is at the synapse of the "vertical" and the "horizontal", the spiritual and the material, the noumenal and the phenomenal, the joined articulation of soul/substance. The Christian symbol, the cross, represents the consciously intentional articulation of soul/substance, of word/work, the "ex opere operantis" dissolution of self in the communal work of natural sacrament. Christian work is identified by love's word, by the intentional articulation of natural sacrament. As institutional religion has been a witting party to the degradation of natural sacrament, it must now function as communal soul, initiating the intentional restoration of natural sacrament—desecrated nature. This it does by enabling family in its social work of conscience building and by reconciling its theological mission with its cosmic rational mission.

Dead on Water

Industrial waste, chemical runoff and spills of raw sewage are ever more frequent causes of catastrophic kills of fish and ground water pollution. Dead fish on dead streams is a metaphor for the religious stagnation of consciousness in our time. The pattern of religious culture, like that of the secular, is mercantile and consumerist. Claim of rights of dominion and exploitation for personal and institutional profit are made on all aspects of nature by systemic religion and politics. The soul of consciousness seems to be dead on the stagnated waters of self-obsession. It is in the nature of consciousness to uplift and not stagnate on a *status quo*. Consciousness is yeast in the bread of life; it is the energy of self-aware matter. It is the intention of Sacrament. Yeast that becomes inactive needs to be replaced with active yeast. Centrism and staticism are old yeast, good only to be thrown out. Informed consciousness of the transformational universe is new yeast having the power to uplift.

Consciousness is the process-energy of the quantum-electric universe. "Quantum-electric" may be understood as a scientific term for "body/soul", in which, body represents the *quantum*, the material, and the *electric* represents the spiritual, the soul.

"Soul" may be thought of as the consciousness of cosmic energy/rationality. Teilhard de Chardin conceived of *rising consciousness* in the evolution (genesis) of the cosmos, of floral/faunal life, of humanoids and of altruistic service (Anthropogenesis, Christogenesis). He envisioned *noosphere* as an aura, a zone of consciousness enveloping and penetrating Earth. Earth's atmosphere is gaseous fluid that interpenetrates the floral webs of land and the animated oceans on Earth's surface. Earth's atmosphere is being understood by science more and more as the vital breath (consciousness) mediating the many interactions of energy and gaseous exchanges of Earth vitality in all its complexity. Earth atmosphere is the wave field aura of spiritual consciousness in which the varied media of communication play out their intricate messaging patterns, functioning patterns of consciousness that are *Sacrament*, that is, formulated in material structures as qualified by their energetic complexity. The "sign" (material structure) conforms to the qualifications of energy (soul—grace), and qualifies transformational potentials.

Earth aura (atmosphere) is a *sign* of Earth intelligence *gracing* all vitality. Self-aware vitality, the intensional construct of body, is essentially transformative, that is, it is interiorly driven by intention to self-perfect. Consciousness cannot countenance staticism, the prison of centrism. It needs an open sphere in which to operate. Lacking a healthful and open sea of communication, consciousness stagnates. Suffocation sets in, and little by little the dead forms of life come to surface and become toxic detritus poisoning the higher-life environment. The enduring hope of life is that transformational consciousness can win out in its determination to rise above stifling toxicity.

Connecting the Dots.

In our childhood most of us probably did dot-to-dot drawings, which pictured objects on paper by connecting dots that were numbered in sequence. I can remember catching on to how they worked, and once I did I would first scan a new piece and get a quick picture in my mind what the object was before I started to connect dots. Well, cosmic rationality uses dot-art logic.

Materiality is a complex layering of spirituality. It is a complex onion of many skins. Materiality includes layered skins of faith, hope and love—the graces of communication, consciousness and conscience—processes of cosmic rationality. Photons are electron-carried "dots" enlightening the art of self-reflective consciousness. Though obfuscated in electron complexity, photons are intuitional contact points that are generously interspersed within the layers of material complexity. The cloudy layers of molecular electrons overlap and interplay. The impulses of interwoven photons impact consciousness and make photochemical connections. Photochemical layering is a mechanism of diversification, of transformation accomplished by electron sharing. Even though the entire materiality of the body is in a continuous process of being replaced, the dot-art of reflective consciousness knows to keep in place the essential points which preserve the continuity of the complex fabric of cosmic rationality, the *hypostasis* of divine/human linkage—the *Christogenesis* of humankind in the *noosphere* of Sacrament.

The connecting bond between faith and rationality (reason), for example, is love, even as love is the bond connecting faith and hope. Thus, the relationship of reason and hope is identity, for both are the prospective insights of consciousness. Love is both cause and effect of the joined growth of faith and hope.

Human self-awareness tends to be preoccupied with self, and, thus, to overlook the inherent fact of the coherently *spiritual* nature of all material. Self-awareness is a glass which is limited in its holding capacity; when it overflows, which it always does, the overflow remains outside the purview of reflective consciousness and escapes conscious connection, unless the glass can be made to increase its holding capacity. Evolutionary transformation is a process that does just that. It brings more of the spiritual layers of consciousness into reflective connection.

Learning to scan in broad perspective the connecting dots in cosmic rationality can be helpful in keeping one's sense of personal relationship in clearer perspective. The jolt of the earthshaking cataclysm of 09.11.01 can be, if we let it, an expansive experience that enlarges our personal glass and lets us come to a saner sense of place in cosmic reality. Perhaps we can become a little less self assertive and a little more sensitive to the fragility of the communal bonds that are too easily trashed by the idolatries of possessiveness and self-obsession.

Sylvester L. Steffen

Serving Whose Truth?

DO THEOLOGIANS, OR BISHOPS FOR THAT MATTER, CARE OR EVEN WONDER WHAT LAY PEOPLE THINK—ABOUT THEM?

What's an ordinary person to think about the public posturing going on by theologians in the matter of their being boxed in by Rome to sign a "mandatum", just to satisfy Rome and the bishops?! Is it another example of angels dancing for position on a pinhead? To put the question in perspective, what does Jesus' preferential option for the poor have to say about this hullabaloo? The "business" of instituted theology is clearly made a "business" matter of institutional Church, that is, a matter of keeping its fences mended for economical and political reasons. After all, if the fences are not tightly maintained the faithful might roam around and put their coins in the collection boxes of other institutional churches. The jealous politics of institutional religion, of building and securing denominational structures, have been obstacles to the nitty-gritty work of religion at the grassroots level. Theology, radically biased toward institutional Church's structural security, misinforms, misleads and misdirects with respect to God's Word spoken to and on behalf of the "poor". Only the poor, and children, are truly free of the adulterating bias of the institutional Church's agenda. The urgency of the poor and the candor of children give insight to Jesus' theology. After all, the "theologians" Jesus picked to be his first apostles/disciples were a rather ragtag lot of hang-around types and would-be fishermen. And even more humbling, Jesus wouldn't recognize them unless they "become like" little children! Like Nicodemus in Jesus' time, today's theologians seem torn between the polar opposites of temple regulations and Jesus' counter-cultural vision supported by a less than impressive crowd of ne'er-do-wells. Nicodemus could muster only enough courage to speak well of Jesus in the security of the dark.

I don't mean to belittle theologians, bishops or Rome. They do play an important role in the scheme of things; however, each should be respectful of the other. But, when in their behavior toward each other they act like selfish, scrappy, bully kids, they need to be told. Don't they realize the harm and scandal caused by their manifest institutional self-conceit? Who will tell them? If they are left to point the finger at each other, only more heat is generated and no resolution is achieved. The message must come from the people. Professional religionists need to come off their high horses and recognize the place of theology in the "natural" scheme. This might shock professional religionists, but it is true, I believe, that the natural goodness in human relationships, say, for example, of a bishop/priest marrying, is less a scandal to people than the Church's excommunication of a married bishop/priest.

Where there is dialog with God, theology happens. Theology cannot be reduced to institutional ideology. Yet, most of us are churched so to think. Theology is not the exclusive prerogative of an elitist group sharing the same train of esoteric thinking about God-human relationships. Theology is no entitlement proprietarily owned by an elect group of Church-diplomaed individuals. You and I, every body, are "theologians", if the example of Jesus' election of leaders and followers says anything at all. Do bishops and theologians, Rome and bishops think that people can't see through their petty, childish power plays? If they do, they are wrong.

The common denominator of community, locally and globally, religious and secular, is family relationship. When family solidarity is faith-full, hope-full and love-full, harmony prevails, and children learn to bring family-acquired lessons to all other relationships. When family is perverted, whether by institution-dominating religion or by greed-dominated secularity, disharmony breaks out and families become dysfunctional. Family trinity, father-mother-child, expresses God's authentic Trinity in nature, the tri-polar rationality that qualifies all other rationality. Father is the faith symbol, the reliable source of spiritual/material insight; child is the hope symbol, the conscious product of father/mother communication; mother is the love symbol, the all-caring soul inspiring family harmony.

Are there irreconcilable differences between bishops and theologians? Shame on the bishops and theologians! Get beyond differences. Resolve them by dialoging on the specifics. Who has the last say? God. Opinions will always be at variance. Give differences a chance to reveal truth. Truth can be several-sided. The opinions of bishops are not superior to theologians, nor, are theologians' opinions superior to bishops'. We're all equal before God. Inferiority, superiority are non-starters. What matters are tolerance, understanding and love, trinitarian bases of social harmony. Are there irreconcilable differences between Rome and bishops? Get beyond them. The Councils of Trent and the First Vatican, to the contrary, Roman Catholicism is not a kingdom or fiefdom belonging to any one, pope, king, college of cardinals, or royal retinue; kingdoms and fiefdoms are unjust and worn-out political structures of the past. Church is a global people struggling to find its common soul and common destiny. Church is, should be communally, common light, common understanding and common love—the origin and destiny of social life on Earth.

Static-world fixations, imposed by engrained habituation on transformational consciousness, are stumbling blocks to religious credibility. Religious consciousness must be brought forward in a way that accommodates new knowledge. "General" and "Special Relativity" in the quantum-electric cosmos is the consciousness that has replaced the ancient belief of earth-human centrism. It is overdue for theology, and Church institutions for that matter, to reconcile old faith to this reality of human-divine relationships.

Reconciling Religion and Science

The professionals of religion and science still feel obliged to react against each other and to defend their old fidelities. Religion and science professionals, beholden for their livelihood and committed to their institutions, continue their wars with each other ever since the inquisitional condemnations of theologian/philosopher Giordano Bruno and scientist Galileo Galilei, whose condemnations occurred only 33 years apart.

The ongoing war of mutual intolerance hurts religion, science, and people, for the public largely accepts evolutionary science while it wants and needs to live by faith. The cultural wars between the institutions of religion and science are deeply unsettling because the hard lines, for example, separating evolution and Scripture's literal account of creation, are put forth in a way that excludes one another. Notwithstanding science's evidence in support of evolution, religious *foundationalism*, in all its stripes, opposes it.

The historical investments of institutional religion and science in holding turf against each other motivate insiders of both disciplines to resist accommodation. To insiders, reconciliation seems tantamount to surrender of turf.

Alienated in self-electionism, institutions wage their wars and establish their own "industrial/military" complexes, which prevent the dilution of their fixated presumptions. Movement by religion toward reconciliation is slow at best if not dead in its tracks. When Pope John Paul II acknowledged that evolution is "more than theory" it was well received, though it was less than a ringing endorsement of evolution.

While it is understandably difficult for the parties to reconcile, they must, for their wars are deeply hurtful. Quantum religion challenges the divisions between religion and science.

Quantum relativity presents nature's original, but groundbreaking *platform* as the venue for science and religion to be reconciled, and urges them to bridge their divide and to heal the schizophrenia they aggravate. Quantum religion is faithful to Catholicism's universal spirituality without being fixated in institutional intolerance. Whether or not these institutions accommodate to one another, quantum religion accommodates Christian Faith and scientific knowledge, and enables people to live in peace with both.

Bishops' Listening Session

Ames, Iowa. August 24, 2001
The National Catholic Rural Life Conference
Statement by: Sylvester L. Steffen

Sylvester L. Steffen

THE "FARM ISSUE" IS AN ISSUE CALLING FOR WHOLE CHURCH ATTENTION; IT IS NOT JUST THE CONCERN OF RURAL PARISHES.

Dear Bishops: shepherding is an issue of land and people. Corporation dominion over the land (global resources) and farmers, without sensitivity for sustaining and restoring natural providence, is the prostitutioned infidelity of instituted greed, which Scripture calls "idolatry". I wish to direct my comments to the magnitude of the "farm issue" in terms of it being a global crisis of social sacrilege enabled by religion's misdirection (irreligion).

Religious fidelity begins with faithfulness to the "purpose of life"; humankind is a self-aware agent in life's purpose. Fidelity to the natural laws of relationships enables the working of natural/divine providence (grace). This law is fundamental to authentic religion. With respect to "Farm Programs", Churches have largely taken a disinterested role in that farm issues seemingly engage an idiosyncratic and shrinking rank of its members. The work of the National Catholic Rural Life Conference is a notable exception to this generalization. However, even the work of NCRLC, as far as parish involvement is concerned, has for the most part been limited to rural parishes. Bishops as a group have largely been uninvolved. It is to be hoped that this is about to change.

It is a fundamental fact that all parishes are "rural" churches in the sense that the natural basis of social, communal life roots essentially in the land, upon whose providence all depend. Therefore, urban, "non-rural" Churches, the People of God in cities, have as much at stake in rural issues, if not more, than farmers themselves. It is a matter of critical importance to the whole Church how government allows and enables corporate exploitation of the land, and the people on the land. It is the responsibility of Church leadership, bishops, to engage the whole Church in this conscionable matter of the whole community.

Multi-national corporatism is neo-colonialism, a "culture of death" imposed on the many for the profit of the few. I personally witnessed with chagrin and frustration the systemic and systematic feudalizing of agriculture in America over the past 30 years—which continues unabated even now—while Churches yet do little more than grieve with displaced farmers, too often at their graves. I commiserated personally with the late and beloved Father Norm White, NCRLC director of the Archdiocese of Dubuque, who was angered and frustrated for the apparent futility of his efforts. His struggle and farmers' struggles have been virtually ignored by Churches—even by rural Iowa Churches! Bishops mistakenly presume that the Farm Bureau, for example, works in farmer interest rather than corporate. The NCRLC, and its lonely directors, struggle on against the odds.

I will keep my point very simple. Please, for God's sake, involve the whole Church in the necessary activist work of Catholic Rural Life, for nothing less than the future well being of humankind is at stake. Make no mistake about this, that corporate exploitation as it is occurring presently is an abortion of the land as well as of the People of God. Future generations will pay the price for our irreligious abuse of vital land. Natural productivity, American farmers and the global community are being reduced to serfdom by raw corporate arrogance, and diversified providence is being massively aborted.

Please bishops! Isn't it quintessentially Jesus' mandate to his Apostles that they inform public conscience and challenge corporate idolatry wherever it rears its ugly head? The ultimate victim of exploitive irreligion is Mother Earth, mothers and children. True worship of God expects in the least that Churches confront cultural irreligion and defend and succor widows and orphans, not contribute to increasing their numbers. It begins in justice, with equal access for all to land and its fruits. Please, become engaged in land and farming issues and engage the whole Church, the global Church. What credibility do Churches expect to have if they are not identified foremost in this primary act of worship? Thank you.

Ethical Land Use
By Monica R. Steffen.

(A College Assignment Essay. Fall 1987)

Sources:
"Land, Theology and the Future" by John Hart, an associate
    professor of theology at Carroll College.
Theology of the Land, 1987, The Liturgical Press, Collegeville,
    Minnesota

"'Adam', that is mankind, has as partner and mate,
'adamah', land. Humankind and land are thus linked in a
covenantal relationship, analogous to the covenantal
relationship between man and woman.
    Unfortunately, in our society we have terribly distorted
relationships between men and women, between 'adam' and
'adamah', distortions that combine promiscuity and
domination, precluding in both cases loyal, freely held
covenantal commitments. Likely we shall not correct one of
these deathly distortions unless we correct them both…we
shall not have fertility until we have justice toward the land
and toward those who depend on the land for life, which
means all brothers and sisters."

Walter Bruggemann

Solutions to the problems concerning the control and development
of land will not be found easily. Be it for better or worse, the
economic philosophy, which has shaped the American attitude toward
land, is wedded to the ideologies of individual rights. Leonard Weber

refers to this dominant American social philosophy as the Traditional Ethic.

For better, it has provided each one of us with the opportunity to concentrate on self in the belief that the best way to contribute to the welfare of all is for each to concentrate on his/her own needs. It was intended to minimize the threat of "big government" by reducing its role to one of a negative function, that of protecting the rights of the individual. But because private interests have failed to make meaningful contributions to the welfare of all, government has been given affirmative functions in forms of social services, environmental protection, and the list is getting longer. Furthermore, for worse, this Traditional Ethic has provided individuals with the opportunity to claim exclusive rights to the land, excluding others so that land can be legally enclosed.

The consequences of the techno-economic philosophy to which we subscribe are far-reaching. They touch and concern our industrialized urban centers where people are treated as interchangeable parts for corporate machinery, and they touch and concern our Mid Western agricultural societies in a similar fashion. As Wendell Berry puts it:

> "The rural community – that is, the land and the people –
> is being degraded in complementary fashion by the specialists'
> tendency to regard the land as a factory and the people as
> spare parts. Or, to put it another way, the rural community is
> being degraded by the fashionable premise that the exclusive
> function of the farmer is production and that his major
> discipline is economics."

Is there truly anything productive about stockpiles of rotting corn, fields that need chemical applications to yield more of the same, or toxic aquifers?

One of the keystones in the Traditional understanding of human rights is the right to possess private property, and with a few restrictions, the right to use that property as the owner sees fit. This social ethic also includes a model of distributive justice, that is, a system of distributing the benefits and burdens of society among the members of society, known to us as the free market system. Proponents of capitalism would suggest that this is the most fit manner of determining who gets what through a system of private exchange; however, it is not a system designed to insure that everyone's material needs are met. Weber further suggests:

"The perception of land as a resource to be used primarily for private profit is a natural corollary of this philosophy…

The land use ethic that contemporary American society has inherited is primarily an economic ethic. The value of the land is determined almost completely by its role in the market system. Land, like any other resource, is worth only what you can get from it. It is worth what you can do with it or perhaps to it; its value is what you can sell it or its products for. In this ethical system, land has value precisely as property. Land is not considered good in and of itself; it is good only if it is good for something. Its value is instrumental, not intrinsic."

This prevailing use of the land has brought us to (perhaps past) the crossroads of crisis. As John Hart suggests in his essay "Land, Theology, and the Future":

"There is a 'crisis of ownership'. The limited lands of the national domain are being gathered into fewer and fewer hands. Increasing numbers of people are being uprooted from the soil and left without a stake in land that was carefully cultivated or freely roamed by their ancestors for generations.

There is a 'crisis of use'. The soil, the forest, the rivers, and the ground itself are exploited with little concern for social or environmental effects. Poisoned air blows across the land, poisoned water flows through seeps into the land, and poisoned soil blankets the land.

There is a 'crisis of values'. In contrast to the best ideals of political founders of the American republic and of the religious founders of the church communities that people the republic, 'greed' has replaced 'need' as the prevailing response to questions of land ownership and use."

Given our position either at the crossroads between life and death or well down the road into the valley of death, what are our options for reconciling our positions with the land and the people of this land? The signing of the INF treaty by the two major super powers should serve as a reminder that it is possible to move in the direction of compromise even with two seemingly opposed ideologies. Although it would be difficult for a free man/woman to endorse the Communist Ethic as an alternative to our Traditional Ethic, some serious consideration must be given to the idea of somehow integrating, adopting or evolving a new ethic that begins to come to terms with the problems we are experiencing with our current economic system of land use management.

Alternative Models.

Various models can be found in some of the different cultures that exist currently in America, although economic pressures are threatening to eliminate their existence. The Mother Earth Ethic was dominant in the Native American culture centuries before the European ideals were transplanted. It embodies three basic principles:

(1) Mother Earth is sacred and cannot be owned
(2) Mother Earth is to be respected and cared for, and
(3) Mother Earth's gifts are to be shared by all living beings through the ages.

Such principles, while being a beautiful statement of truth, would be difficult to re-establish in a system that advocates the ownership of private property.

A Family Farm Ethic found a home in America when European peoples immigrated to these rural regions almost a century and a half ago. This ethic viewed land as something to be closely connected to, not in a Mother Earth sense, but rather in a pragmatic, anthropocentric way. Much of the plight of the family farmer rests in the fact that this connectedness was not dominated by economic considerations although the public policies that focus on agriculture do have that bias.

Models of land use have also been proposed throughout history. These models have had little if any success in finding a way into the arena of public policy, but their ideas are well worth mentioning.

Perhaps one of the oldest of these land use ethics is found in one of the most widely read texts in existence although it is probably not recognized as being a land use alternative by a good majority of its readers. As Walter Bruggemann suggests in his essay "Land: Fertility and Justice", the Biblical Covenant is based on a theory of land.

"The fundamental dream of Israel is about land. Israel is a social, theological experiment in alternative land management...The core tradition is intended to promote an alternative to the imperial system of land known both in the Egyptian empire and in the Canaanite city states."

The imperial system referred to here was one where land was regarded as a tradable commodity: "land became an arena for commercialism and all the social problems that emerge when the strong are aligned against the weak." (Sound familiar?) In Israel's theory of land, the land is assigned to the entire community as a gift, a trust, an inheritance, in other words, the connection between the social unit and the land was inalienable and contained within it a sense of continuity.

A more contemporary model of land use, the Ecological Ethic, is the one Aldo for which Leopold argued. The key focus of this ethic is upon the land itself and on the relationship we need to have with our life-giving environment. Leopold argued for a sense of responsibility toward the natural environment which has for too long been considered as no more than "mere property". He expounds on this ethic with these often-cited lines:

"All ethics so far evolved rest upon a single premise: that the individual is a member of a community of interdependent parts...The land ethic simply enlarges the boundaries of the community to include soils, waters, plants, and animals, or collectively the land."

He further elaborates on one of its underlying principles:

"A thing is right when it tends to preserve the integrity, stability, and beauty of the biotic community. It is wrong when it tends otherwise."

Respect for nature and the land is the petition being made. Respect implies striving for appropriateness in the uses to which the land is being subjected. Sara Ebenreck in her essay "A Partnership Farmland Ethic" suggests substituting the phrase "working with" rather than "using" so that partnership with the land is immediately implied. According to Ebenreck, there are three principles, which apply to this partnership; they require:

(1) respect for the fundamental nature of the land
(2) use that does not destroy that nature
(3) returning something of value in exchange for the use.

It stands to reason that knowing the land is essential to acquiring respect for the land. Since there is a certain amount of variability in soil, water and nutrients from site to site, appropriateness in use would appear to be a regional issue, something that is site specific. While scientific communities, advocate this ethic particularly by those in the life sciences, it has not, as a system of ethics, been adopted by the general public. It probably influenced to some degree, the environmental legislation of the 1970s.

The last land ethic use model this author would like to set forth is one, which Weber calls the Communitarian Ethic. Central to the belief of Communitarianism is the notion that individuals and human rights need to be understood within their social context and that the responsibility of each individual is to contribute to the common good. Individuality finds its meaning and identity within the constructs of the human family.

The notion of social responsibility in terms of community obligations is something that has received little recognition in our American intellectual heritage. Liberty particularly and equality have received attention, but the community ideal has not had much impact. The concept of community revolves around cooperation, solidarity and commitment to common goals. Social responsibility requires individuals to be "aware of" and "concerned about" the impact of their individual actions on others in society. What the Communitarian Ethic would like to suggest is that we try to promote those activities, which would be beneficial to the lives of others, rather than to pursue our own economic self-interests. This ethic would definitely have an impact on business practices.

Kenneth Mason, a former president of Quaker Oats, had a very appropriate attitude when he said,

> "Making a profit is no more the business of a corporation than getting enough to eat is the purpose of life."

The social contract with business as it exists today is in fact one that is purely profit-motivated. This position is no longer adequate. There is a need in this day and age to come up with a new definition for this concept of profit. Daniel Lufkin has suggested that it be "one that will address corporate gains and losses, not only in terms of dollars, but also in terms of social benefits realized."

The development of land would also be impacted by such an ethic. Land ownership involves significant responsibility in a system of Communitarian Ethics. In his essay "land Ethics: Toward a Covenantal Model", William Everett identifies four specific ownership rights:

(1) the right to "use", meaning occupancy or "neutral" exploitation which does not alter its continuing use
(2) the right of "income", referring to the fruits of the property
(3) the right of "transfer", which means either selling, giving or leaving the land as a legacy
(4) the right of "alteration", meaning development.

Alteration or development, according to Weber, is least defensible as a "private right". He suggests that "social responsibility requires that we recognize that development rights are limited by human needs and the needs of the environment." This is not to imply that development is altogether wrong; it simply means that development also has a social responsibility to the community in which it occurs.

As far as "human rights" are concerned, the proponents of a Communitarian Ethic would argue that we need to establish a theory of "basic human rights", that is, what is necessary for one to live a minimally dignified human life. The focus of such an approach to the concept of "rights" would be to redirect orientation from what we can get for ourselves to what are our social obligations in providing what is needed for essential human dignity.

There are clearly some conflicts that exist between our well-established Traditional Ethic and some of the alternative models, which were proposed, particularly the Communitarian Ethic. In looking at the issue of positive human rights, there seems to be a question of priorities that needs to be addressed in terms of what role government should play. I personally feel, as I am sure many do, that this is not a government responsibility: to suggest that they assume the role of assuring that everyone's essential needs are met would be asking for the trouble that comes with "big government", and I am sure that in the long run this would prove to be ineffective. The individuals must take such responsibilities in a given community, by the individual community, and perhaps by the churches in those communities. Introducing a role for community churches to play in a system of government that maintains a hard line separation between church and state is a risky notion, but if every denomination within the community were to set aside their doctrinal differences in order to work for justice and equity within their respective communities, I believe the threat of any one religion emerging as a power could be maintained.

The role of private agreements, particularly covenants, should be reviewed as a potentially effective tool within smaller communities for enforcing regulations pertaining to land use. They must, however, be carefully scrutinized for the potential to restrict according to an arbitrary personal bias. It is recognized that complex problems arise in trying to enforce covenantal agreements, which are created by developers to meet the conditions of their personal plans for an area. Perhaps the concept of "community covenants" tailored to the specific needs of a region or community would be useful. I would suggest, however, that first the issue of contract language needs some serious attention.

One "disease" that is rapidly spreading in our increasingly specialized, technological world is something called "technical illiteracy"; no matter that nearly one third of this great nation is functionally illiterate, we insist on creating amongst ourselves profession-specific languages that confuse and confound the real issues at stake. Corporate America and, unfortunately, the intellectual community are, in effect, building yet another "Tower of Babel". It is urgent that we try to re-establish communications that seek to inform each other rather than alienate or differentiate one "species" professional from another.

And now, for some unpopular comments about the architectural profession: The current trend toward historic preservation in architecture and landscape architecture is problematic of all the ailments we all suffer in this technological era. From my limited knowledge of Biblical history, it appears that we are doing for architectural history what Deuteronomy did for the Old Testament, which is basically to plea for the return of the forgotten system of beliefs but with a much stronger voice. It serves to remind the few who do understand that language to our rooted-ness to the past, but after all, it is no more than that, a reminder of a time now past. Meanwhile, development and its architects continue to erect technological icons that deify the economic force to which they aspire. I should not be so severe in my assessment as to include all architects and developers, for I am sure that there are some who do have legitimate concerns about how to effectively design for the communities they serve.

But now to the real issue at hand—the mission of Iowa State University, which was by design created to serve he architectural region and its communities. As we move into the twenty-first century and toward an intellectual perspective, it would be selfish to assume that the contributions of a university be limited to the sights of a particular region; however, Iowa State University must still recognize and act on its responsibilities to the communities that make its very existence possible.

While it is recognized from the outset that this land grant university is indebted to the corporations which provide the monetary fuel to keep the "light" burning here, care must be taken so as not to create a camp of indoctrination that mass produces the interchangeable parts for the functioning of their machinery. What it should seek to do in educating its youth is to inspire the individual with a sense of confidence and personal mission so that he/she will have the courage to present creative solutions to the problems faced by the communities in which he/she serves.

Before taking a look into a future mission for this university, it will serve us well to take a look at its past intent:

It was under the Morrill Act of 1862 that public lands were made available for the establishment of institutions that would teach subjects relating to agriculture and mechanical arts. There were two major adaptations where American universities are concerned that influenced the whole pattern of their development:

(1) the transfer of the government of a university from faculty members to a lay board of trustees. That Iowa State should have a "Board of Regents" seems to indicate an administrative trend towards imperialism. I would advise that they take a close look at why they are, in fact, serving on this board for it is important that they recognize the "trust" placed in them.

(2) democratization of the curriculum. In the 19th Century, farmers sensed a need for education in order to face the challenges of an expanding and changing rural population. Their educational needs were not those of a scholarly elite, but rather those, which were broad, liberal and practical in nature. What evolved was a type of Land Grant University that taught agriculture and veterinary science along with some traditional subjects like home economics, journalism and engineering.

Land grant colleges developed another feature; the idea that the university was a service agency for the entire community led to the development of state experiment stations and extension services.

So the original mission of the land grant university stems from the needs in agriculture into the notion of community outreach. As I see it, that mission still exists; perhaps what needs to evolve is the sense and direction of community outreach (rural, urban and international).

Recognizing that the intent of the current administration is to streamline the focus of this university, it must be warned of the dangers of too much specialization. From an economic standpoint, if you produce too much of a particular thing (be it technicians for computerized operations, business administrators, or corn) you run the risk of super-saturating the market. And when that happens things have a tendency to fall. If it is really the intent of this university to establish a unique mission, I would suggest that it seriously consider a theme directed on ethics, not so much in terms of human morality, but rather one centered on idealized and realized land use.

I believe there is some real potential in the notion of developing a strong community and regional planning curriculum here which strives to combine the scientific research geared toward creating humane environments with the technological and artistic know-how of the architectural disciplines.

The time is critical...we are at the threshold of breaching the sustainable limits of the land and with our current programs of outreach, geared to force unsound production of homogenized commodities, we are all destined to fall. We must strive to liberate the creative potential of the individual; we must resist the temptations of becoming enslaved by our technology and by the economic principles that drive it. If we are all to be masters of our own destinies, we must search for identity within ourselves and within our communities.

[Note: On June 9, 2001, Monica Steffen died of a brain tumor, which was diagnosed on February 26, 1997. In her living Monica practiced her philosophy. She crafted her own college curriculum of study, which gave her professional skill in many areas: plant and animal sciences, horticulture, art, drafting, architecture, music, languages, poetry, and others. She became especially beloved to many in the New Hampton, Iowa, community for her detailed

knowledge of plant cultivars, regionally suited, and her unflagging readiness to discuss and to draw professionally landscaping plans for private citizens.]

Delusional Terror

In the normal course of events parents convey to their children a sense of right from wrong and the difference between actions that are mindless and done "on purpose". In petty grievances between peers, an aggrieved party may seek recourse from his/her mother, and if mother's ear is closed to the aggrieved pleading, the aggrieved may plead more urgently, "But mother, he did it <u>on purpose</u>!" Such urgency has a better chance of getting mother's attention. Appeals made to heaven for wrongs done "on purpose" call for urgent response.

And so it is with societal misdirection. Wrongdoing done out of ignorance is less culpable than knowing actions done purposely. Intentional motive distinguishes malicious actions from poor judgment. Repetitious and knowing wrongdoing is a special kind of malice that demands redress, especially when it is systemically ingrained and pernicious in consequence. The wrongdoing of ingrained patriarchy is a case in point. Globally, mothers need to rise up against patriarchy's pernicious malice and demand, "No more!"

The face of delusional terror is male, it is systemic and it has overtones of religious pretense. Just because it roots in the deep past and has enjoyed cultural sanction does not excuse its culpability. Personal and collective conscience now knows its malice, which cries to high heaven. Intentionally and unintentionally, the immediate victims of delusional male terror are women and children. Genetically coded in deep evolutionary history, aggression is the one thing that males are most certain about—aggression in presuming a superior self-righteousness and in the self-assertion of presumptive certainties.

Male aggression invades theologies and houses of religion no less than other venues; and alienated by their own doing from sensitivity for femininity, males become even more deluded in their presumptions. In the Seventeenth Century, the intended targets of male terror were women whom males determined were witches in league with devils; in our time, calculated perpetrators of violence are Arabic *Al Qaida* and Islamic Taliban. In Judeo-Christian-Islamic cultures women have systemically been repressed by formal alienation from leadership roles in Church and State. A new, rational theology is needed that recognizes the unique leadership charisms of women that are essential for societal well-being.

Violence is a natural phenomenon in the transformational universe: lightning, tornadoes, hurricanes, floods, earthquakes, volcanoes, etc. These are events of energetic disturbances outside of human cause and control; however, by their thoughtlessness and abuse of natural resources, for example, denuding mountainsides of their tree cover, humans can exacerbate the violence of floods, winds, etc.

Delusional terror is of a whole different kind than nature's terror. It is violence that people inflict on each other but which may be avoided. It has some connection with the fact that life forms are in nature food for one another. Instincts are hardwired, causing animals to prey on other life. Dogs chase rabbits; cats stalk mice and birds. Humans are heir to the same hardwiring. In their preying instincts humans target their search for food on every accessible resource, whether on land, in the air or sea. Human success in invading all these habitats is threatening to land life, sea life and birds; now frightening even to humans themselves because of species' extinctions and mortal havoc to essential webs of life. Humans inflict violence on each other in asserting exclusive right and claim to consumable resources in the territory they dominate. This gets to the very root of human violence on humans, for economic and moral force are used to assert exclusive claims and to exploit resources even though such claim and exploitation may work against the common welfare.

Delusional terror originates also in sources that otherwise are legitimate when fairly exercised. For example, the enforcement of law, when it is harshly exacted to the letter without consideration for mitigating circumstances, can be a source of public terror; insistence on "truth", when it is exacted specifically and literally according to some institutional definition and without sensitivity to human relationships, can be a most tragic form of terror for it presumes its righteousness on God's authority.

The heavy hands and arrogant minds of domination are of ancient lineage. They are the tools and the calculus of kingdom builders. They belong to the likes of the Roman Caesars, the European Czars, and even the Corsicans, Napoleon and Giovanni Mastai-Feretti (Pope Pius IX). Grand delusions fired their zealotry.

Arrogance and domination are the cultured preference of the all-male types, which manifests, for example, in male-exclusive organizations such as the Roman Catholic hierarchy, mixed lay and clerical groups such as the Knights of Columbus and Opus Dei. These jealously hold to the all-male godhead of absolutist, linear theology. These are remnant cults of centrist doctrine and the centrist worldview, whose time has come and gone; neither now well serves God, man or nature. Conscience requires human consciousness to rise to a new level. Quantum religion may provide the insight necessary to raise consciousness and conscience to a new level of understanding and doing, and humans to a new "phylum of love" and a new level of civil relationship.

The rational (eclectic) person is one who is open to evolution and determinism, which may effectively collaborate to adopt and codify what promotes communal and personal welfare. In the course of evolutionary change, few things are genetically codified in such a way that they become uncontestable absolutes. The pattern of contesting and proving values of actions, and the merits of beliefs and practices, is perhaps the nearest thing to an absolute; and that pattern is the Trinitarian process of rationality, namely, of communication, consciousness and conscience, and its (their) spiritual products of faith, hope and love. The psychological process of reason is *with physical/psychical outcome, and morality is the outcome of the valuation of process*.

State and Church have through history functioned as agencies of terror. Institutional fundamentalism (literalism), in interpreting law and in defining and applying "truth", has forever been a source of human terror, and probably always will be. However, Church and State are both morally obliged to act in humankind's common interest, and both need together to seek out better understanding and regard for the lot of common vitality. The global appetites of human masses are already so overwhelming to nature that nature's resourcefulness is at risk of being shutdown as a reliable life-web for humankind.

Institutionalized arrogance is an enormous obstacle to public-interest government and religion. The vested interests of bureaucratic hierarchies interfere with their fair-minded work on behalf of the public. The institutionalized creeds and structures of religion make rigid claims on truth and fixate institutions in inauthentic mindsets, which prevent them from evolving into people-sensitive agencies.

Judaic based religious traditions, present day Judaism, Christian and Islamic Churches, equally have violently inclined prejudices that allow extremists to foment the kind of terror that led to the violent take-down of the World Trade Center in New York City on 9-11-01. This event is a consciousness-shattering event that calls for a more thoughtful response than violence-for-violence. Nothing less is needed than a global rethinking of resource distribution and fairness by corporations toward the marginalized. Multinational corporations need to assess their global role in marginalization, disease, poverty, and violence.

This isn't the place to detail the complex roots of today's global predicament that date from the 1400s. An excellent source of insight on the origins of today's global mischief is a well detailed work of Hugh Trevor-Roper, "The Crisis of the Seventeenth Century, Religion, the Reformation and Social Change", 1999, Liberty Fund, Inc., 8335 Allison Pointe Trail, Suite 300, Indianapolis, IN 46250-1684.

Though institutional religion may age and get tired, God does not. In modern times the soul of Christianity has grown out of its threads and is in earnest search for new cloth. Staticism, centrism and absolutism are worn-out threads that no longer provide cover. In increasing numbers people associate these *isms* with historical despotism, the discredited monarchies of Church and State.

Hope springs eternal because in each generation new consciousness is born. In each newborn, consciousness is put to the anvil of trial-and-error. The confines of consciousness are challenged by awareness of past misdirection and by the universal aspiration of greater well-being. The consciousness of global self-awareness now seems more cognizant of its earthly co-dependencies in the conjoined webs of natural complexity, perhaps because people largely acknowledge and accept themselves as spiritually/materially originated in the continuity of an expanding universe.

And where is God in a cosmos that is expanding in ways yet to be determined? As good an answer as consciousness can give is that God is inherent in creation, still and ever in process. In this consciousness God is neither denied nor diminished. To the contrary, the mystery of the divine deepens as the cosmic mysteries deepen. It only gets better.

The consciousness of mind in the universe is like yeast in rising bread. Consciousness is the interiority, the intention of sacrament. Consciousness in matter traces back to original Big Bang energy. Consciousness serves as an ever-growing tide of ascendancy that is accustomed to frustrations. The cosmic energy of transformation compels change, change that is initiated in trial-and-error bases of proven redundancies. Consciousness is reflectively engaged in pursuing escapes from frustration. And so, global humanity is not without conscious resources by which it can escape the terrors of its own making.

In the biblical Tower of Babel account God is described as being chagrined over the fact that humankind had come to be of one mind, one language. The mono-cultural mentality of corporate farming has harnessed natural diversity and wealth to its singular "tower-building" intent of profit and control, an arrogant enterprise that pierces the sky and enters the very house of God. But God frustrated this pursuit by causing the ancient tower to collapse and by confusing people with languages of different purpose.

The expansion of colonialism was fueled by the common expectation of European nationalities to extend themselves into unknown lands and to appropriate the wealth of unknown peoples. The singular mind and language of this expectation is consumerism— the prostitution of global Earthlife for personal and national advantage. Indigenes threatened by outside exploitation are right to stand up against violations of their own persons and property. Crass consumerism is a repetitious and singular mindset and language that ever since the sin that consumed the fruit of the "middletree" in Eden still possesses humankind and still frustrates divinity.

America can surely do better than carry the torch for Old World vices. To colonially depressed countries, America represents the amalgam of Garden of Eden instincts and European expansionism, but also hope for something better. The Western World shouldn't be surprised when flare-ups of hatred and violence occur in response to old habits of domination and control of global resources.

In secret moments of frustration and pain we may sense within our own selves the stirring of our own proclivity for violence. Violence is perpetuated in lives conditioned by violence. Violence and terrorism are dark proclivities toward mean destruction and human hurt. Potential terrorism is a seed lying dormant in every heart. In an environment of terror the seed is almost certain to sprout and grow. The global prevalence of violence is perhaps now unprecedented in prior history. Violence affirms the negative imprint of passion, but passion does not have to be negative, destructive. Jesus taught us the example of positive passion. Passion for violence produces violence and ruin; passion for non-violence in the face of violence mitigates the proclivities toward terrorism and ruin.

The process of rationality can detect and avoid consciousness-caused terrorism. By the conscionable determination (will) of consciousness, violence and its precipitous outcomes can be obviated. Fixations in belief of selective, divine *chosenism* can predispose consciousness toward delusions that may justify acts of terror "in God's name". Fixation in chosenism may incline one to arrogate his religious rationality over that of others, especially if the faith of the other conflicts. Violence in this case is a product of a hyped zealotry.

In some way or other, terror always seems to be a byproduct of polarity. Terrorism and absolutism seem to be bedfellows who have in common a radical fixation in divine righteousness; when opposing ideologies are fueled by a like radicalism, there seems to be no rationality that can stop their determination to exterminate each other.

Christianity preaches a starkly different rationality. For injury suffered it teaches a response of love. The preferred rational way of dealing with anti-social behavior is to remove people from the circumstances that provoke violent response. The elimination of causes, which provoke violence, may prevent it, but reaction to violence with violence will not.

Deeply Tri-Atomic

The internal tension of the tri-atomic molecules, water (H-O-H), and carbon dioxide, (O-C-O), is primed to attenuate wave energy in the infrared band of the electromagnetic spectrum. The chlorophyll in green plant cells opportunely appropriates light (photons), digests water and carbon dioxide, and structures them into the light-laced "glycogen agency" (C-H-O-H)—Earth-life's singular, energetic confection and structural resource.

Isaiah realized that "all flesh is grass", that human consciousness is self-aware bread. Jesus brought to humanity the consciousness of divinity in grass/flesh. Flesh is grass "transubstantiated". The substantive incorporation (transubstantiation) of grass in higher life forms by consumption and digestion is the exchange medium of divinity, the mystery work of flourishing vitality.

Wheat is word transubstantiated. Bread is word transubstantiated. And, as Jesus is the Bread of Truth, so are we called to be. In the seed's dying, transubstantiation is enabled. We are bread born from and cast upon the water—broadcast word *right as grain*, even though *green as grass*. Soul and body's sustaining growth is not possible except communally empowered; body/soul's empowering food is grain-truth continuity energized in essential soul/substance relativity.

Natural spirituality is wave-field empowerment, cosmic energy distributed in wave-field consciousness and overlapping molecular complexities. Conscious wave-field complexity is driven internally and externally, internally by wave-energy of electron momentum, and externally by the electromagnetic wave-stimulation of impacting photons. Elemental "sexual" ambiguity grounds in the tension energy of the positive atomic nucleus and the "negatively" energetic electrons. Electrical ambivalence, positive and negative potentials, is the motor-energy driving all Earth-life transformation.

The empowering food of the body, which is at the same time the empowering food of soul, is basic glucose whose glycosidic linkages trans-substantively provide structural energy/substance from which vital edifications arise and diversify. The house that God built for the people is a *house of bread*—Bethlehem—the edification of Divine Providence in natural providence.

Soul and body, we are "bread", communal edifications of bread. As bread substances, we are necessarily "agents of bread", conscionable doers, whose lifework is the provisioning of bread, the care taking of soul/body. In provisioning bread and in grain keeping, we brother-keep and we sustain community. Jesus' words over bread at his last supper, "This is my body", are spoken with an insight of cosmic consciousness intended to enlighten our own substantiating sense of communal connection. This priesthood, the provisioning of material/spiritual *Eucharist*, the breaking of bread, conscionably and mutually obliges man and woman alike. We are obliged to empower each other in this universal priesthood, not alienate others from it.

Sun Power. Light itself is a floodtide of particles (photons) whose quantum energy is defined by particulate wavelength. The "substance" of light is real and ever active in the networks of air, water and soil, which collaboratively assemble the stuff of life; human life is part and parcel of Earth-life's intertwined networks.

The universe's coherent sense of physical order is inherent in the laws that predispose subatomic particles (quanta) to their electrical agencies as induction motors that purposefully drive atoms and molecules.

The purposeful symmetries of universal substances (particulate quanta) occur through attractive and repulsive accommodations; atoms are elementally formed by the communicational power of cosmic dialectics. The nucleus of the atom carries a positive centripetal potential, while the electrons in the nuclear skies carry negative centrifugal potentials; their purposeful exchange is life's language of structural logic.

The solar flux of photoelectric energy vitalizes the Earth and cocoons our privileged orb in diversifying networks. The golden sun is and will always be Earthlife's singular source of sustainable, revitalizing energy. If we humans conduct ourselves in ways consistent with preserving the global network of sunshine-structured life, we may recover the conscionable awareness that Sun Power is the original and only sustaining life-source of global ecology and economy.

As individuals and as communities of individuals, we human beings become fulfilled, both physically and spiritually, when we self-accommodate within life's sustainable networks. Because we are network agencies within networks, we have the frightening ability to trash and crash other networks, but also the vitalizing ability to strengthen and amplify the waves and ways of diversity.

Grain power is sun power. Particulate quanta flow from the sun in waves of radiant energy, graduated in a broad spectrum of high intensity to lower intensities. Life on earth is complexly diversified because of the unique ability of system networks to capture the sun's radiant substance. This trapping of photoelectric substance is the original and only sustainable method of harnessing nuclear energy. The seed, a life-incorporated storage battery, is simply stored sun power—the food of life's network web—the substance of flesh. Conscious of this natural value, humans have from times immemorial recognized the unified religious/secular value of cereal grain (grass) seeds in the commerce of life, the *eucharistic sacrament* of communal sustainability—creation's living connection with its Creator.

The destiny of the live plant is directed toward the purpose of producing seeds, and the destiny of seeds toward the purpose of producing the live plant. In this cyclical strategy of nature, energy and matter return to renewed usages, over and over again. In addition to re-using nature's structured symmetries, living matter is augmented by the green-cell theft of solar substances (photons). By this substance acquisition from the Sun, the substance and diversification of life on Earth are augmented.

The seamless blanket of life, which covers Earth, that is, ocean life, soil life and airborne life, seeks to proliferate and to diversify in interdependent ways. The common denominator in life's diversified, sustainable networks is the sharing of solar energy. Sensible practices that are true to life are enabled by a working knowledge of the mechanisms and processes essential to sustainable living, that is, the intentional corroboration of agri-culture with natural sustainability. Sustainable life-culture is the foundation of sustainable society. True agriculture is intentional *sacrament* working. We should be shocked and chagrined over corporate abuse of land and of farmers, and for the lack of sensitivity in the exploitation of land and life.

In coming to a more informed understanding of the natural commerce of corn grain we may better understand natural/divine providence in human affairs. Einstein's equation of Special Relativity speaks directly to the natural economy of living energy, to the substantive effects of sunlight, and specifically, to the electromagnetic synthesis of living reserves of photoelectric food energy. Earth-life's critical defense against entropy is the seed. Without the seed there is no diversified life, no photoelectric reserve by which to diversify and sustain life.

The molecular microcosm. The way in which sun power is captured is incredibly clever. It is accomplished in the microcosmic world of atoms and molecules. Each atom has a nucleus, which is at the center of its micro "planetary system". Traveling in orbitals (skies) around the nucleus are electrons, which gain impetus when exposed to the stimulation of harmonic electromagnetic radiation. Atoms are joined and become molecules when the electrical imbalances between them are accommodated by electron exchange/sharing. The molecular transformations that occur when electrons are shared or surrendered create subtle new potentials in the electrical fields of the new molecules. Pathways involved in the exchange of electrons are subject to qualifications by the subatomic components in the nucleus and in the electrons. When atoms are clustered in a particular molecular configuration, they manifest (as a unit) a specific "vibrational frequency", which qualifies the interactive potentials of molecules and atoms in the cluster.

Resonance and energy attenuation pertain to harmonic vibration. So characteristic is harmonic vibration to all molecular phenomena that it is reasonable to presume its deeper relevance in even more basic cosmic activities. Favor is gaining for a *string theory of everything* and for a preference of envisioning first substance not as an elemental dot, a *graviton,* but as *resonance strings* that orchestrate cosmic flux at the ultimately least state—some mass-less dimension. String theorists envision more dimensions in first super-strings than can most people. According to latest count, super-string dimensions are up to eleven. [Tom Armstrong, "Elegant Universe: the Zen of Superstrings", Winter 2002, <u>EarthLight,</u> The Magazine of Spiritual Ecology, 111 Fairmount Ave., Oakland, CA 94611, pg 43, a book review of Brian Greene's "The Elegant Universe: Super-strings, Hidden Dimensions, and the Quest for the Ultimate Theory", 2000, Vintage Books] The strings are "...confined within an individual unit of space...a unit of measure so short that if a hydrogen atom was the size of the known universe...a unit of measure...would be no longer than a mature oak tree".

Harmonic (vibrational) frequency attenuates (captures) the energy of compatible wavelength frequencies from the electromagnetic spectrum. This capacity for attenuating (trapping) compatible wavelengths of solar radiation is called a "harmonic" response. The energy state of a molecule, its agency-potential, increases by virtue of its harmonic attenuation of solar radiation. For example, water vapor is in a more highly excited energy state than liquid water. The continuous exposure of molecules to specific wavelengths of radiation may so excite them as to cause them to disassociate from some molecules and to associate with others (ecstatic resonance).

Living systems have the ability to use the vibrational energy of (super-strings?) molecules as well as their structural components. Living systems are in perpetual states of transformation. Life is a transitional complexity of cyclical events involving purposeful molecular arrangements that provide for food acquisition and utilization. By some incredible economy, living systems can chemically rearrange raw substances as they use the energy contained in them, and capitalize on these materials not only for structuring and restructuring their own systems but also for using and storing energy for later use—all of which happens at the atomic/molecular level.

Life itself is a complex expression of wave-field energies subtly disposed in organized matter, whether in the molecule of a human body or in the seeds of plants. The life-regulating ribbons of DNA and RNA dispose cell substances to the autotrophic (self-feeding) work of life's spiral ascent. DNA helices, contained in every cell, store and access the workable codes of living relationships, which have been perfected over eons of time. DNA is an open *computer chip* that enables nature/nurture to adapt purposefully to the contexts of the times.

The continuum of universal transformations, characterized by purposeful redundancies in atoms and molecules and enabled by water, light and the natural laws of continuity inherent in DNA, confirms emphatically the connectedness and unity of life.

Water, and quantum agency. The natural laws of living relationships are indelibly written in/with/by water. The openness of life is a credit to water's transparency. The providence of life (in accessing food-energy) is the doing of water, which is both agent and catalyst in cell-dynamics. Life's origins were enabled by special sets of circumstances in which structural substances and energies were transformed by electron exchange. Earth-life began in the medium of water and totally depends on water.

Hydrogen is present in water in vast quantities, but the recovery of hydrogen atoms from water requires a large amount of energy. Photosynthetic autotrophs (organisms that produce their own light-made food) have devised an elegantly organized system of pigments (chloroplasts), which facilitate the capture of the sun's light energy. They use the sun's radiant energy to raise electrons to excited states and then trap them in such a way that a portion of their energies is extracted in a usable form before the electrons return to their ground states. What is essentially a light-induced separation of electrical charges is sufficiently energetic to split water into hydrogen and oxygen gases. It is also sufficient to generate ATP (an ester derivative that supplies the necessary energy of cellular processes), which, along with oxygen is required to bind the carbon and oxygen of carbon dioxide into biologically useful compounds with higher reserves of free energy potential.

Photons are the energy agents that are intercepted in the chloroplasts of green plant cells, and are used in the assembly of elemental carbohydrates (C-H-O-H), which are the basic food and structural materials of diversified life. Chloroplasts possess light potential responsiveness to photon activation, and therefore, have the electrical ability to mediate the chemical interactivity of water and carbon in the seed. Photons may be operative in the visible as well as in the non-visible electromagnetic spectrum. In plant seed development, including "after-ripening", the contribution of photons is molecularly substantive.

Photosynthesis. Water (H-O-H) and carbon dioxide (O-C-O) are molecules that are found in the air; together they are the raw materials of photosynthesis. Reduced to their essential elements, water and carbon dioxide are the food materials of all life, the original substance of life. During the process of photosynthesis, carbon, hydrogen and oxygen are rearranged into carbohydrate molecules from the elemental originals, which is accompanied by the release of oxygen (O-O) into the atmosphere. Photosynthesis essentially generates two vital results: it splits water and carbon dioxide, which enables the use of their components in the manufacture of glucose (glycosidic molecules), and it releases oxygen into the atmosphere.

The process of photosynthesis occurs in two stages: the stages of *light reaction* and the *dark reaction*. The hydrogen necessary for reduction in the synthesis of glucose comes from NADPH (nicotinamide adenine dinucleotide phosphate). This carrier molecule, along with ATP (adenosine triphosphate), provides the free energy needed to synthesize glucose. During the light reactions, the chlorophylls, enzymes and cytochromes in the chloroplasts of plant cells trap light energy to manufacture NADPH and ATP. These accumulate during exposure to sunlight. In the dark reactions, they interact with carbon dioxide to produce the glucose molecules that power the growth and differentiation of the plant. These are the original substances that form the seed.

$$n(O-C-O) + n(H-O-H) + \text{radiant energy} = n(C-H-O-H) + n(O-O)$$

This is where life's continuity and all food begin, where the edification of living networks begins, and by which, they are sustained. All life depends upon the process of photosynthesis for energy and structure. Humans victimize themselves when they trash this network reality. There is no substitute source for human life!

Chemical Synthesis. The chemistry of life is the chemistry of water and carbon dioxide. Carbon (C), hydrogen (H), oxygen (O) and nitrogen (N) are the basic construction materials of all biochemistry. Water is not only the provider of hydrogen and oxygen atoms to the alchemy of life, but is the biological medium of all cell function and is the energetic environment which enables the chemistry of life.

When the seed has matured and no longer needs the plant water for the transport of materials from the plant to the seed, excess water can begin to be evacuated. The elimination of water occurs at the point where the kernel attaches to the cob; it is through this conduit (wick) that moisture first entered the developing seed and through which it is also released. Unless grain allows for the evacuation of excess water, the ingredients of the seed would be exposed to biodegradation.

When the parent corn plant no longer produces *glycosides* (sugar-based) products, processes that stabilize the seed's chemistry begin in earnest. The movement of water may now reverse its direction, i.e., the excess water, in its free state, may begin to exit from the seed. This stage of maturity is indicated by the formation of the *hilar* cap, a dark membrane at the seed's tip. The individual seed now functions as an independent organism—a living system in its own right.

The new organism's first major task is to consolidate its accumulated food materials. The build-up of stored energy in the seed (carbohydrates, proteins and lipids) may be likened to "charging the battery", for energy potential is stored and stabilized in these macromolecules. More than 85% of the seed is food stored in large starch molecules. These macromolecules are end products of a long succession of glycosidic linkages that assemble stable, carbon-based molecular chains. The condensation process of carbohydrates separates out water components, the hydroxyl (OH) and hydrogen (H) radicals. At each new linkage in the chain, a molecule of water is released for re-use and may eventually, after many transformational usages, be cycled out of the linked food system.

Glycosidic complexes. Carbohydrates constitute the most abundant fuel source available to living plant systems; sugars and starch comprise the primary food components, and cellulose, the primary structural component.

Glucose is the basic plant and animal food that is produced by green plants and stored in seeds. Simple sugars are converted into polysaccharides (complex glycosidic chains) and ultimately into long-branched starch chains in the mature kernel. Glycosidic linkages occur with the release of water from the sites where the glucose rings are joined. The site of linkage is critical in determining whether a particular chain of glucose molecules becomes plant food material (starch) or plant structure (cellulose).

By way of chemical condensation, starch molecules assume a stable configuration that endures for as long as favorable environmental conditions allow. Tighter chemical density generally means greater stability.

Chemical Hydrolysis. The embryo cannot use the complex starch compounds as a food source until they are loosened through the process of hydrolysis. Hydrolysis is the conversion of stored starch to sugars, the conversion of stored protein to simpler proteins, and the conversion of fats to fatty acids. This process involves the loosening of chemical bonds through the addition of water. The more complex the original compound, the more water molecules need to be restructured back into the chemistry in order to alter the macromolecule to simpler, more usable chemical states.

Hydrolysis is the reverse of synthesis. Just as a molecule of water (HOH) is given off at each linkage site in the consolidation of complex molecules, so a molecule of water is restructured back into the condensed molecules and is reconstituted into molecularly less complex constituents. To do this, the OH and the H radicals (water components) must be returned. To facilitate this process, a certain amount of water must be accessible to the seed for restructuring back into the kernel's chemistry.

Whenever a molecule of water is put back into the seed's carbohydrate system, the "dry" matter of the kernel is increased by the molecular weight of the hydrogen and oxygen molecules that are reintroduced into the macromolecules. These reintroduced components of water are no longer water. And providing that these new products are not immediately consumed, the marketable "dry weight" of the grain may actually increase.

Generally, conditions favoring germination are also conditions under which hydrolysis occurs; these conditions generally provide for an increased uptake of oxygen and increased respiration. In the seed, however, it may be possible to have hydrolysis occur even while dormancy restrains respiration. Under controlled storage conditions, hydrolysis may allow for increased accessibility of food stores and increased weight in grain. This may come about through the utilization of residual seed moisture as the stored starch and stored protein convert to simpler sugars and proteins. It is important to recognize that hydrogen and oxygen (OH and H radicals) are real and important contributors to grain values and should not be forcibly extracted from the kernel's chemistry.

Soil, air, light and rain make it right with grain. The power and process of grain comes from the continuity of natural laws. Natural vitality is the *divine grace* of Earth's living network. The energetic means that make grain are the same that make grain useful. Grain is part and parcel of nature's transformational web.

The natural script of grain's genetic coding is a revelation of provident nature. Water's true scripting of grain is a light scripture that belongs to all life. In obeying nature's scripting, humans get it right with nature, with God. Biological wisdom advises that we must *do right* by grain in order to *be right* with God.

"Ecology" is the *home environment* where all life joins in communication. It is the living network into which we are born and upon which we depend. We share this uncommon home with all life in common. There is peace in this shared existence when we individually recognize our place and when we preserve the harmony that provides for all cohabiting life. In this home, corn shares a vital place with us.

"Corn" is a metaphor for the Holy Other that gives being to our very self. Corn is the "grass" of Isaiah; it is more than a metaphor, it is the Sacrament of Provident Reality for land life—the Eucharist of our own being, doing and having.

In the common "home" of the living, nature has apportioned the orderly means by which all are bound interdependently, sustainably. Because humans possess a greater consciousness of the intimate workings of this network interdependency, they can choose either to exploit this Sacred Order for disproportionate self-advantage or to accommodate personal self-advantage to the proportional economy that extends nature's distributive harmonies to All Other. When we conscientiously conform our personal being, doing and having to the cybernetic economy of natural providence, we are conscientiously "communal" and we provide a sustainable future also for our own kind.

The "culture of life"—agriculture—is an obligation beholding all human beings. Metaphor and reality, "corn economy" is the conscionable act of *breaking bread* in a sustainably providential manner. In our care and reverence in the use of living grain, we may learn to become more aware of how to care for Earth's network life and for our own communal and personal welfare.

Public accountability. The cumulus of grain research suggests that the science of *seed after-ripening* needs to be researched for its specific public interest values (energy conservation and grain value enhancement) and for its sustainable economies of on-the-farm management of stored grain. In a real sense, however, it already has been researched and documented over a period of more than four decades in the practices of farmers who have experienced the comparative values of corn when it is kiln-dried (prematurely desiccated and killed) and when it is dried seed-sensitively at atmospheric temperatures.

As farmers changed from ear-corn harvesting to field-shelling, agricultural engineers presumed that the fast drying of corn grain with ventilation fans and kiln heaters was an acceptable solution; the damage inflicted on corn grain was assumed to be negligible. However, user experience has since the 1950s documented serious and costly disadvantages: major investments in short-lived, hazardous, labor-intensive drying and handling equipment; escalating costs for liquid propane gas; explosive dust; crumbly and light-weight grain, lacking in palatability and nutrition.

Clearly, grain that cures (dries) under natural conditions acquires/retains certain food/market values that are lost with fast drying and exposure to intensive heat. In 1908, the Russian scientist Vasiliev already suggested this when he expressed his dismay that "seed ripening" had been so little studied; to him it was clear that at a fundamental level, seed ripening deals "essentially with the synthesis of organic substances." [H.S. McKee, "Structure and Synthesis of Protoplasm, "Growth and Differentiation in Plants: A Monograph of the American Society of Plant Physiologists", Editor, W.E. Loomis, Iowa State College Press, Ames, Iowa, 1953, p. 328.]

It is time for the United States Department of Agriculture to take seriously seed-sensitive drying and the farmer-managed storage of corn, and to become advocate for ecologically sustainable economies that serve the public interest, rather than energy-consuming and polluting interests. That this value has not been researched is admitted in the November 17, 1981, letter from the U.S. Department of Agriculture (Larry M. Seitz, Research Chemist for the North Central Region's Science and Education Administration at the U.S. Grain Marketing Research Laboratory, U.S.D.A., Manhattan, Kansas) to Sylvester L. Steffen. "...I don't have any data that would either support or negate your claims regarding an increase in dry matter during storage of corn when moisture content is 15-20% and temperature about 40oF." [See APPENDIX C, Cereal Chemistry Seminar, Pg. 275]

Sustainable vitality is Earth's most active complexity equilibrating interactive, intensional forces. Vital economy is the sustaining balance of supply/demand, a natural interdependency that generates ecologic networks and provides internally for co-dependent membership. The human body, for example, is a communal unity of untold billions of symbiotic organisms contributing collaboratively to the intensional economy of physiological processing. Human beings are just one of many intensioned networks within Earth-life's continuity of networks. Network disintegration occurs when critical masses of some network complexes spin out-of-control and ravage bigger networks in which they obtain. Cybernetic sense should work to restrain disintegrations inflicted by run-away complexes. "Success" is equilibrium-security in life's network; "failure" is the collapse of equilibrium. Salvation, health, and sanity are conscionably secured in network equilibrium and continuity. Our generation is witness to the unparalleled proliferation of network disintegrations, catastrophes of global proportions—unconscionable, humanly induced interventions that trash the essential continuity and equilibrium of network life.

The human predicament consists in this, that ignorance trashes cybernetic sense while arrogance crashes it. Is it possible for exploitation run-amok to be restrained? When life-webs overload network resources to the point of breaking down ecologic generators, the essential resources of "Eden's Middle Tree" are themselves terminally threatened. Cultures that put terminal demands on network resources are in real terms "cultures of death". In the *culture of death*, we westerners are the most destructive of intentional agents because our tools are the most powerful ever, and our ignorance and arrogance are sanctioned, professionalized, absolutized, and idolized by *religious* culture.

Global resources/sources are being imploded by the egoistic exploitation of consumer demands. The image that comes to mind—of the egoistic catastrophe now happening—is of Earth as an animal carcass crawling with unaware maggots. When the last remnant of life on the carcass is totally consumed, the maggots are gone with it.

This specter, of Earth-life's networks collapsing, raises the question: "how far are global human populations from the ultimate predicament of their own self-destruction?"

(Excerpted and edited from "The House of Bread", by Sylvester and Monica R. Steffen, *Tenth Book* of the **Poetree** trilogies, self-published by Sylvester Steffen, 1999)

Wheat as Word
"All flesh is grass". (Is 40: 6)

Word! That capitally intensive
Work of brain!
Entropic enterprise of thinking, speaking, writing!
Nay! Gut-originated inspiration.
The work of photosynthetic grain,
Of water and carbon from the solar sky
Spectrally melded!
Basic glucose! Transformative gold,
Naturally concocted for life! Not for profit!
Dialectical corn!
Marketably packaged!
Spatially. Politically.
Industrially, exploitively idolized
Over systems symbiotic
And holistically economic.
A fabricated and cruel abuse!
An irreligious excuse
To choke Earth in a strangling web
Of fraudulent technology,
Ecologically ignorant of essential biology.
The body is bread; bread is body;
Abuse of one abuses the other.

Sylvester L. Steffen

Gingerbread Jesus

Because God's ways of self-revelation in nature are untiring, religion should never be tiring or tiresome. Tiresome religion needs to be retired and renewed if it is to be a vehicle of self-fulfillment.

In concert, Albert Einstein, Joseph Campbell and Teilhard de Chardin, orchestrate for us a new harmony of perennial wisdom advanced in contemporary knowledge, a collective vision that enlightens the Christian insight of Word-made-flesh, the Way, the Truth and the Light. The vision they advance is one of relational (religious) experience that is rooted in quantum cosmology, quantum philosophy/psychology, and quantum theology.

Einstein, Campbell and Chardin might be characterized as the prophetic trio of cosmic rationality; Einstein, the prophet of cosmic *communication*; Campbell, the prophet of cosmic *consciousness*; and Chardin, the prophet of cosmic *conscience*. The harmony of these is a *trimorphic resonance* that might open all people, all religions, not just to greater tolerance but also to greater collaboration toward universal civility. Nothing short of such collaboration is adequate to bring humankind to personal and communal *pleroma*. The insight of cosmic rationality is the insight of *spirit and truth*, the means by which all might be "born again" as Jesus explained to Nicodemus, the Temple minister. Birth *in spirit and truth* is the means by which the "kingdom within" is edified. The collective tide of inspired truth affords a spiritual baptism of universal cleansing (redemption) and whole making (salvation).

Einstein enlightens consciousness in the cosmic understanding of the photosynthetic physics edifying all life. Einstein puts in place for us an understanding of the photoelectric nature of quantum energy in his *Theory of Special Relativity*—the equivalency of energy and matter. The energy (E) of mass is MC^2, where C represents the speed of light. Further, light is understood to be both particle and wave. Einstein's universe is transformational because it is quantum-electric. This is as true of the human body as it is of celestial bodies, though they change in their individual ways. Einstein's *Theory of General*

Relativity tells that bodies *attract* (or repel) one another because of their electromagnetic, gravitational characterization.

Campbell enlightens us in the open significance of *word* (metaphor) and its capacity for amplifying spiritual consciousness. Campbell compellingly informs us in the perennial power of metaphor (myths) to expand insights of personal and relational consciousness in the connectedness of contemporary individuality with the continuum of vital experience. This is passed on from generation to generation in stories, which engage the imagination (image-making) in consciousness-raising.

Chardin enlightens us in the consciousness that spiritual energy (soul) is a subtlety of cosmic energy. Chardin leads us to understand that spiritual, psychic energy is cosmic energy highly complexified in molecular structures and advancing in *Christogenesis*.

The intentional and compassionate interpenetration and remaking of the spiritual and the material is the universal experience of *Eucharist,* the continuing edification of soul and body. In graham crackers and gingerbread cookies, infants and children (adults too) experience material/spiritual self-edification—Eucharist. As adults we should be able to grow into a deeper sense of divine presence and communication without losing our sense of appreciation and wonder of the divine at work in sweet bread. St. Paul reminds us that children are expected to speak, think and act as children, but when they become adults they are expected to grow beyond childish speaking, thinking and acting. As children we learn literally, in pictorial detail, but as we gain in knowledge, our understandings move beyond the mere literal. This is true of *religious* knowledge as of other knowledge. If this weren't true, novels, poetry, etc., would be absolutely pointless. So it is dead-minded of institutional religion to expect adults to believe literally, simplistically in the manner of a child. Faith that is fixed in childish literalism is *fideism*, the dependent consciousness of children.

Human beings of all times experience the same progression of life's stages, the same spiritual awakening of life processing from the fertilized egg through life to death. Because every life journey is unique, and in its uniqueness "epic" in its own right, it can in its uniqueness be recorded in a way that enriches others as they struggle through the particular trials of their lives.

In reading biographies, novels, poetry, etc., we can associate experiences in our own spiritual pilgrimage with others about whom we read even though they differ in details. The conscious linkage of spiritual discovery and progression, enriched by the narratives of the living of others, enlightens and authenticates *religious* understanding and behavior.

In the building of personal character we challenge ourselves with the bigger-than-life examples of people who have gone before us. We do this individually and collectively in institutionally ratified culture. By doing this we seek to transport the seeming insignificance of our individual lives into the bigness of heroic living, which is illuminated in the lives of others. Thus, the religious cult of Christianity, its commemorative ritual and doctrines, which affirm the exemplary living and teaching of Jesus Christ, help us put our lives in harmony with the hero experience of Godlikeness in Jesus' word and work.

The power of religious ritual and doctrine is the power of metaphor, a transcendent consciousness that challenges our spiritual imagination to make room for all the connotative dimensions that are presented by recalling the specific words and works of others; these constitute idealism, heroism and soul-stretching potential.

The equating of divinity with *bread*—the power of divinity in bread to edify body and soul—is, for example, a sustaining metaphor of infinite *virtue*, power. "Myth" is an extended metaphor, a life-describing graphic that includes storied accounts of heroic living and teaching. The power of myth, like the power of metaphor, is the power of insight that enables personal consciousness to lift oneself in mind and fact above ephemeral drudgery and find in it the raw material for inspiration and heroic living.

For example, "Jesus is gingerbread" is a metaphor, which when taken literally leaves a time-stopped pictorial image in the mind. The power of the metaphor is not the time-stopped picture but is the spiritual development of the connotations associated with Jesus and gingerbread. When exaggerated attention is given to the time-stopped mental picture of Jesus as Gingerbread Man, and the putting of the picture in cinematographic motion (transformational) is neglected, fixation on the picture results and the point of the larger spiritual message is missed. The value of the pictured moment in our own lives is enriched by a sense of place in the "big picture", which is not static.

Just as the lives of other people, recognized in the many identities shared with our own lives, are measures against which we gage and understand ourselves, so metaphors (myths, stories, biographies) are reflective collections of common lifetime experiences concretized in event-pictures that capture for our edification the lessons of experiences that flow through our own event-full lives. The concrete manifestations of events vary in specifics according to the qualifying contingencies of each lifetime, but are, nevertheless, universally recognizable in the events of every lifetime—sort of like a movie that is played over and over again with a perseverating harmony that resonates in our perceptual consciousness; the sequences of personal living repeat but vary in the specifics. It is helpful for us to be aware of life's gene-patterned cycles even though we don't really know them until we live through them. Metaphor facilitates imagination in anticipation of the sequencing of life's cycles. Metaphors are aids that enable us to connect psychically, spiritually, in advance of living through the cycles of our own life. By anticipation we may be better prepared to pass through life's cycles with equilibrium even in the face of unexpected deviations.

Quite obviously we can only live our own lives; thus, it is not only not helpful but it may be disruptive if we objectify and fixate metaphors in literal time-frames and thereby miss the point of their enabling grace that comes from the recognition of ourselves as their subject. Life "in general" is really *nonbeing* for every life is particular and individual; in the particular lives of others we can self-identify. In the process of self-identification we may recognize every life as a metaphor whose subject I am. By such identification we might become more sensitive and tolerant for it brings home the truism that what we do to others we do to ourselves. As an extended metaphor every life is full of lessons of truth. The truth in metaphor and myth is not in literal detail so much as it is in powerful insight that connects meaning and experience. If mystery is imprisoned by fixation in details it loses power to raise consciousness. As Joseph Campbell says, "we diminish the experience of its real depth". Necessarily, metaphor and myth occur in the dimensions of time and place, nevertheless, their real universe is consciousness, the realm of the spiritual, as Jesus affirmed, "The kingdom of God is within". And that is a mind-expanding recognition.

Because religion, relational consciousness, comes to us over time in the sequence of "picture frames" of learning and experience, we easily fall into a misreading of institutional religion and respond to it in selectively fixated, time-stopped frames of reference. However, this reading of institutional fixation isn't just a misreading for it does have bases in fact. The root of institutional disconnectedness in our time from the public mind (sensus communis) roots in institutional fixations that seem irrelevant to modern consciousness. Fixations blur the conscious sense of personal place in cosmic continuity and the revelation of the unbroken connection of all vitality. The fact of cultured disconnection raises important questions that are quite well raised by Joseph Campbell:

"How in the contemporary period, can we evoke the imagery that communicates the most profound and most richly developed sense of experiencing life? These images must point past themselves to that ultimate truth which must be told: that life does not have any one absolutely fixed meaning. These images must point past all meanings given, beyond all definitions and relationships *to that really ineffable mystery that is just existence, the being of our selves and of our world.* (Emphasis added.) If we give that mystery an exact meaning we diminish the experience of its real depth. But when a poet carries the mind into a context of meanings and then pitches it past those, one knows that marvelous rapture that comes from going past all categories of definition. Here we sense the function of metaphor that allows us to make a journey we could not otherwise make, past all categories of definition." (Joseph Campbell, "Thou That Art, Transforming Religious Metaphor", edited by Eugene Kennedy, 2001, The Joseph Campbell Foundation, New Way Library, 14 Pamaron Way, Novato, CA 94949, pp. 8 & 9.)

Except for metaphor, communication is unimaginative, drab, heavy and unexpansive. Metaphor projects personal consciousness into new dimensions; it projects on the colorless definitions of one's individual life the colorful experiences of other lives not unlike our own. Through metaphor we consciously become what we experience, which is what authentic religion also helps us do. Christianity, for example, would have us project the life of Jesus onto our own so that we might *be/become Jesus*. So strong is the Christian consciousness of divinity experienced in Jesus that each newborn is recognized as a *Christ of Second Coming*. The life of every newborn is destined to be a progression of humanly common stages that lead through open gates into continuous, transformational experience, not unlike the transformational stages of Jesus' life. If by fixations we "stop at the gate" as we come to it, we fail to grow authentically into the next dimension and instead our consciousness becomes fixated in a "room" that we need to leave behind. If we dwell too long in the past we may become permanently traumatized by it and fail to experience life's usual dimensions.

Individually, we are ships riding the tide of cosmic consciousness. Transformation experience teaches the lesson of "the non-constant constant", namely, that in the next moment we are different than we are at this moment. The "constant" is the fact of change, which causes us not to remain the same, the "non-constant". "Being" is not a *state* rather it is a captured image of oneself caught in a stage of *becoming*. *Being* is but a snapshot of the process of self-*becoming*. When I look at a snapshot of myself I recognize that it is not I, it is a representation of a being that was but which is now *nonbeing*. This is the mystical realization of cosmic necessity, of essential transformation. We are not now the static being, the image that we see in a picture rather the picture is a fixation of being that is now nonbeing.

Harmony, like wisdom and religion, is in the connectedness of becoming, not in the fixation of being. It is also in the nature of consciousness, of spirituality *to become*, to transform, to process from the ground of static being to growth in the experiences of interdependency. We are by cosmic necessity a spiritual "resurrection" from the natural ground of prior, common being. All life is an extension of the "cosmic *monad*". (As used here "monad" means the prior *singularity* that gave rise to the Big Bang and the aftermath of the continuously expanding universe to which Earth and we belong.) The personal experience of transformation, of resurrection, is the *good news of rising consciousness*. In the tide of rising consciousness we all experience resurrection. The rising tide of personal consciousness swells the tide of cosmic rationality, of communal consciousness. Because of universal connectedness, the covenantal obligation to inform and uplift personal consciousness serves to prevent the stagnation of the waters of common spirituality.

Consciousness is the spirituality of quantum-electric self-reflectivity seeking conscience; it is yeast fermenting wheat-flour bread in the process of becoming. The tide of *rising consciousness* is metaphor and reality; it is the change of *transubstantiation*—the reconstruction of substances into new materiality/spirituality. We *become* "bread" to one another; we are conscious agents of transubstantiation. *Resurrection* is conscious spirit rising; *ascendance* is the interpenetration of personal consciousness; *transcendence* is the intergenerational continuity of spirituality/materiality. Resurrection, ascendance and transcendence are a tide of spiritual awareness that proceeds from generation to generation and transforms individual life into newness.

Like Jesus, we are gingerbread. Every child should be told the story of the Gingerbread Man for its spiritual implications. It is a fun story as well as a morality story. The Gingerbread Man is Everyman. Everyman has a Maker. The child becomes Everyman. Everyman experiences the thrills and chills of childhood, of life. Everyman faces trials, failures and accomplishments. Everyman is a prize and a price to every other man. Everyman struggles to discover himself and to self-accommodate to life's circumstances. In life as in death, Everyman is a Gingerbread Man that others seek to consume. And like the Gingerbread Man, Everyman does not ultimately escape being "eaten up". The lesson of the Gingerbread Man is like that of other mythologies. The theme of every life, of every mythology, plays out in repetitive patterns but in details that are unique to individual experience. In this reality is prophetic vision. The lesson of life, though lived out in different details of fact, is not the fact-details, but is the spiritual edification we discover in the living examples of others.

Resurrection spiritually consists of new consciousness becoming fleshed-out in the intentional living of every newborn. The spiritual "monad" is a unity, the collective consciousness, cosmic rationality, that arises and renews in the life of Everyman. (Obviously, Everyman is also Everywoman.) The spiritual monad is the conscious continuum of cosmic rationality, ever arising, ascending and transcending by means of the resonant agency of the trimorphic processes of communication, consciousness, and conscience, and, by which Eucharist, communion, the communal virtues of faith, hope and love, are newly fleshed in each generation. Love is both the ultimate beneficiary and benefactor of trimorphic resonance for it alone endures as the end state of resonance processing, the sustaining consciousness of reality.

The Mass is the central celebration of Catholic Religion. It commemorates the essential act of every Christian life, the act of self-giving to other. Jesus' total *pouring out* of self for others exemplifies how we are *Eucharist* to one another. In Eucharist, we are, like Jesus, *bread*. We are future-life's connection to Earth. This is a spiritual reality that drives human relationships. Eucharist is cosmic rationality, the essential self-giving of transformational reality. The process of Eucharistic living is a moment-by-moment fact event, the sustaining continuum of divine presence. The past is nonbeing; the future is nonbeing; only the present is being-in-process. This is the Old Testament recognition of God Presence—JAHWEH—*I Am Who Am*. In personal and communal reality, transformational living is the Mass, the commemorative celebration of God Present—the sacrament of cosmic continuity.

Mass, in its large context—The Big Picture—celebrates the transformational universe, the cosmic continuum. It is for us humans to celebrate it consciously, intentionally, unselfishly. Cosmic necessity implicates us in transformational sacrament, in the death of self for the life of other.

Sylvester L. Steffen

Equal Opportunity Sin

In reflecting on religious history one might conclude that the "god-game" is at least the second oldest profession. Like human nature, prostitution is a phenomenon in two dimensions: selling soul (spirituality) and selling body (materiality); in its theological guises, the first may be less obvious, but it is prostitution nonetheless. Whether in the form of pimping "god-talk" or pimping sex, prostitution is profiteering at the expense of others. In the exploitation of others, both are ugly. When analyzed from the perspective of evildoing, profiteering religion is more pernicious in its consequences. The peddlers of pretentious religion know they need to create a devil worse than themselves against whom to sell their product, and sex seems to be their devil of preference. Caveat emptor! Be wary of religion that targets sex as the worst sin and evil. Let's be honest. Humankind can very well get along without pretentious religion, but it cannot get along without sex. There is no decency in the perversion of religion or sex; but when high-road peddlers threaten fire and brimstone for the sins of the flesh, hang on to your pocketbooks!

It is a serious sin to trivialize words and to torque consciousness, for by so doing God is put to the test. All manner of social havoc comes from it. Tampering with the personal consciousness of others is high-road prostitution. Spiritual pimping on the high road of pretentious rationality scorns the material instincts of low road pimping. Which one really deserves the greater scorn? Prostitution is an equal opportunity sin.

The high-road merchants assume the right to make God more imposing and foreboding than God Almighty does Her/Himself. As to the extent in which Divine Presence interjects the Divine Self in natural affairs, God does it just about right. Human response is right when it is sensitized to God Presence in quantum relationships, and when it comports itself with fidelity to God-scripted nature. Professional theology has contributed to muddling the message of God-scripted nature and our place within it. A quantum perspective on relationships might help change our theological vision, our relationships with nature and each other, if we seriously engage it.

The universal call of "quantum religion" is a voice within expecting fidelity. *Quantum religion* attends, moment-by-moment, to the intimate word/work of Divine Presence involving each individually. When fidelity flourishes, so do hope and love. Tune your senses to fidelity and experience hope and love. Spread the Word.

An Alternative to Violence
(The Dialectic of Love)

Radical love is the most reasonable and the most effective solution to violence. Love must be a personal habit of mind and behavior before it can have effect on others. A saint, an ordinary fellow, who lived in the turbulent circumstances of the Middle Ages, comes before us today with a way by which each of us can practice radical love and harmonize interpersonal conflicts.

St. Francis of Assisi, born 1182, Giovanni Francesco Bernardone, was the son of a well-to-do cloth merchant and a mother of noble heritage. Opportunities as a prelate or moneyed merchant were before him. In his youth (1202) he got caught up in a provincial war between Assisi and Perugia. He was captured and imprisoned for a year. From his war experience he saw first hand the destruction of churches and he experienced a strong inner call to repair God's ruined churches. Unfortuitously, but not unimaginatively, he loaded one of his father's horses with as much cloth from his father's inventory as he could. He proceeded to sell the cloth and the horse to raise money, which got him in trouble but didn't get him far in his endeavor to restore churches. Francis opted an alternative of renouncing material possessions, family ties, and committing himself to a life of service to the needy.

Prayer is mental/vocal dialog that seeks to conform personal disposition to divine purposes. The prayer of St. Francis is a universal mantra that may best serve humanity in search of its true self. The prayer of St. Francis reveals the disposition of mind that became the hallmark of his life, as it was for Jesus. Collectively and personally, there is no better way by which to overcome terror than the adoption of Francis' mindset and way of life. The repetition of his prayer might do wonders in our own lives and in others.

Prayer of St. Francis

Lord,
Make me an instrument of your peace.
Where there is hatred, let me sow love,
Where there is injury, pardon,
Where there is doubt, faith,
Where there is despair, hope,
Where there is darkness, light,
And where there is sadness, joy.

O Divine Master,
Grant that I may not so much seek
To be consoled as to console,
To be understood as to understand,
To be loved as to love,
For it is in giving that we receive,
It is in pardoning that we are pardoned,
And it is in dying
That we are born to eternal life.

Appendix A.

Religion: A Rational Consideration
Of the Basic Relationship of Creation to the Creator
(This paper was written to satisfy a Theology Course Requirement)

February 1957
Sylvester L. Steffen

Stating the Issue

Scope of the Term.
The term "religion" is a word, which indicates man's relation to God. The fullness of this relationship cannot be realized unless it includes *man* in the fullness of *his* nature, namely, as a being composed of body and soul [male and female]. In the hierarchy of being spirit is superior to matter, hence, spirit should receive first considerations, not, however, to the neglect of matter. The primary faculties of spiritual beings are the faculties of intellection and volition, hence, the activities of these faculties must be seen in their relation to religion. It is our purpose to analyze religion in the light of the foregoing in order to arrive at a full appreciation of the role it plays in the life of man.

Meaning of the Term.
The word religion is derived from the Latin, either from, "religare" (to bind – namely to God), or from "religere" (to take carefully in consideration, to ponder over, to weigh conscientiously and reflect upon with due care – especially that which is divine and holy). (1). However, an adequately complete concept of the word's meaning is arrived at only by combining [the meanings of] these two words. In the word "religare" is indicated the free action of the will in choosing to bind itself to the Supreme Good; while the word "religere" indicates that this act of the will is founded on the rational considerations of the intellect. Thus, an act of religion, though

consummated by the will, must, to be in accord with man's nature, be based on convictions of his intellect.

> "By means of this virtue we honor the Lord our God inasmuch as we acknowledge and proclaim His greatness, majesty and dominion over us, and at the same time confess our own littleness, lowliness and dependence on Him. Religion, consequently, includes in itself two requisites: first, lively acknowledgement of His infinite perfection and dignity; and then, humble subjection to His infinite power and dominion." (2). [Emphasis added].

The Natural Virtue of Religion.
For the present, our discussion will refer to religion as a natural, acquired virtue. By "virtue" we mean a *habit perfecting a certain faculty to its proper functioning*; and by "habit" we mean with St. Thomas *a quality inhering within the intellect and will, ordering these faculties to the attainment of their proper ends*. The object of the intellect is knowledge of beings, which includes the essences of things in their relation to true order, and the realization of one's place in that order. The object of the will is "the good" as it is proposed by the intellect, which ultimately must be the "Supreme Good". In the realization of its Ultimate "Object" the spiritual being must elicit acts of adoration, the necessary act of religion. (4).

Method of Procedure.
Realizing then, that a rational understanding of our relationship to God is basic to any bona fide understanding of religion, we will proceed to make a rational approach to at least some of the fundamental aspects of the relationship of creation to God; first, Angels, then the material order, an-organic and organic, man—the composite of spirit and matter, then Christ, the God-man, and finally, again man in view of his new relation with God because of the Incarnation. When we come to where Redemption enters our discussion religion will then be considered as an infused supernatural virtue.

The Hierarchy of Dependence in Creation.

Principle.

A work that lacks continuity is the work of an inferior intellect. Hence, in creation, we should expect to find a [pattern of] continuity, nowhere else paralleled, because of the infinitely perfect Intellect that produced it. Our approach is intended to bring out the marvelous continuity of creation.

Angels.

First in the hierarchy of created beings are Angels – pure spirits. Each is a species of its own, a sufficient entity in itself with dependence on God alone, hence, personal, individual obligations to God. Thus, each in its own inimitable way mirrors some perfection of God. (5). They possess intellects and free wills, their sole faculties, which in their "natural" order cannot err. (6). Hence, in the question of their fall, the answer must be found in an order superior to their own, the supernatural. St. Thomas says the fall of the Angels consists in this, that:

"[T]hey preferred their natural glory in its isolation to the community of the supernatural charity…It is pride because they elected excellency without reference to the more excellent good; it is rebellion because the will of God was that they should accept the supernatural". (7).

By this act the bad Angels severed any hope of attaining their intended relationship with God, a fact they know but their choice remains forever unchangeable. (8). With the "Good" Angels, on the contrary, their happy relationship with God is forever happily sealed; their intellects and their wills have attained their destined ends. Their relationship is perfectly ordered.

Material Creation.

An-organic Nature.

Having considered the most sublime of created beings, we will now descend to the other extreme of the hierarchical order of creation. An-organic "beings", composed of matter only (as opposed to spirit – no reference to *hylomorphism* here) evidently cannot exercise virtue as we have defined it; yet, they have a definite relationship and dependence on God. Their excellence is derived from the fact that in their existence and natural activity, they mirror the intelligence of their Creator. A given element, in a given circumstance, will always react in a definite manner. This uniform activity of the individual elements, as well as the uniform interactions of different elements, mirrors the ordered intellect of God. Ultimately, material beings have their dependence on God, yet in the order of things they are dependent on each other. This mutual interdependence is intrinsic to their very nature. The fact that they do act upon each other is evident and can be easily demonstrated, but just WHY they do so we cannot fully comprehend. All we can say is, "it is their *property* to act so". From their consistent actions [reactions] we can learn something of their nature. By observation and experiment we find evident that some activity is intrinsic to the nature of matter, and this activity is effected only by the mutual interactions of different substances; hence, we can define a principle, that: "a *social relationship*, i.e., mutual interdependence in their activity, is intrinsic to the nature of material beings".

Organic Nature.

A further and more marvelous indication of the ordered intelligence of God can be derived from the continuity of an-organic and organic beings. The division between organic and an-organic in the last analysis is hard to determine. For all practical purposes, it is the division between living and non-living beings; in highly developed living beings the division is evident, but when it gets to the point where the living being is in its simplest form, i.e., the stage of transition from non-living to vegetative to sentient life, it is not too apparent. In fact, if it were proved that all life, animal and vegetative, is nothing more than matter highly organized by intricate interaction of involved physical and chemical activity, the theologian need not get excited (provided the spirituality and immortality of the human soul is not tampered with). Even sensation and feeling are subject to the organism and thus share in the general determinism of matter. (9). It is clearly evident that organic beings are wholly dependent upon an-organic matter for their existence and activity; thus, such a concept as suggested in no way does violence to the activity of God in creation, on the contrary, it bears out all the more strikingly the continuity of creation! The infinite fruitfulness of the divine Intelligence is more fittingly portrayed, if matter, in view of the intrinsic potentialities God has planted in it, would evolve in an orderly and patient process due to its naturally inherent qualities, than if God by separate creative acts should produce a hierarchy of material beings. In the latter case, the continuity of creation would not be as striking as it would be if simple matter were capable of interaction. Nor would the material part of man's nature be excluded from this process, else, the finality of matter as it is striving for realization in its noblest form would be frustrated.

The preceding discussion of spiritual beings and material beings in their relation to God and each other, has been rather extensively discussed, because a thorough understanding of them has been considered basic to an understanding of the full nature of man's dependence.

Man the Composite.
Man is a being composed of both spirit and matter. As a spiritual being he possesses faculties of intellect and will; and thus, as an individual is bound to exercise the virtue of religion in his relation to God. As a being of matter, he is also bound to exercise a social relationship with other creatures. Then, as one being, composed of spirit and matter, man's exercise of religion must correspond to his nature, thus, it must be interior and exterior. (10). Material beings, as we have said, of themselves cannot exercise the virtue of religion, but are by their very nature, wholly determined by the Creator, and thus must honor Him by realizing, in their activity, the end for which He created them. Spiritual beings, on the other hand, see their

relationship to God and are free either to serve Him or not. Thus, in the free choice of the spiritual being, God is much more honored than He is by the necessary, determined activity of material beings. But in man, matter receives the ability of sharing with spirit the free choice to honor God. Hence, by participation, matter can in the proper sense exercise the virtue of religion. Herein the continuity of creation is further expressed; man is the bond of creation wedding spirit and matter. He becomes the sole mediator of matter because he shares in its nature. (11). Here matter is raised to an eminence above that proper to its nature. (12).

Man's Break with God.

But if matter received the capacity of sharing the privileges of spiritual beings, it should also share in the activity of the spiritual being in achieving these privileges. Thus, the test given to man to determine his status with God should be one fully accommodated to his nature, effecting body and soul. At the same time, the nature of this trial should be proportionate to the gravity of its effect (13) hence, it should affect the primary faculties of man's soul, his intellect and will, and the primary faculty of his body. Which is the primary faculty of his body? Is it the faculty of self-preservation or the faculty for the preservation of the species? It seems the latter, since by it is realized more fully the social nature of man, whereas, the drive of self-preservation looks more to man as an individual being. Also, we see the grave social effect the first sin had on man, namely, that all are born with it; which would indicate that it should be a social act, capable in its very nature, of accounting for its grave social effect upon the whole group. Such an act is that by which the very existence and growth of the social body is assured.

The Break Repaired.

If man had existed in the order of pure nature only, he would have had the virtue of religion only as a naturally acquired virtue; but, as a matter of fact, he existed from the beginning gifted with the infused supernatural virtue of religion, and was possessor of supernatural and preternatural prerogatives. (14). With his first sin, man lost his supernatural and preternatural gifts, and his infused supernatural virtue of religion became inefficacious in aiding him to acquire his destined supernatural end. Unless God intervened his fate and that of all his children, [they] would be [in] such [a condition] that they could never rise above their purely natural state; they could never attain that supernatural state [which] God intended for them. (15). Here then, God's plan of Redemption enters the scene. In the light of Redemption, man's whole relationship with God takes on a new character; so, an understanding of religion must include the consideration of man in his status as a "redeemed being".

The Supernatural Virtue of Religion.

After the fall, God must bestow upon each man the supernatural virtue of religion if he is to realize his supernatural end. Religion, then, is a moral virtue directing man in his relation to God. (16). The nature of this relationship is conditioned by man's nature; thus, in virtue of his spiritual nature man has obligations to God as an individual; in virtue of his material nature, he has social or community obligations to God. (17). Because of the social nature of man's obligations to God, each individual has thus religious obligations to his social group, which means, that man's relation to his social group will also determine his relationship with God. Hence, it must be that the social group, "qua societas", has obligations to God. And we have said that religion directs man in his relation to God, and to do this it must be efficacious. From the foregoing we can now derive a definition of religion as a supernatural, infused virtue. *"Religion is an efficacious virtue ordering the life of man, as an individual and social being, in his relation to God, and ordering the society of man in its relation to God."* In the final clause of the definition is meant that religion determines and directs the external act of society by which it acknowledges its supreme dependence on God; in the final analysis it is nothing other than Divine Revelation.

Sylvester L. Steffen

The Manifestation of Religion.

We have sufficiently covered religion as an individual obligation, now we must deal with it as it is to be manifested in the social group. To do this we will consider man's obligation to the social group and that of the social group to God.

First, man in his obligation to the social group: that man has social obligations is evident, first, from his nature as a material being, and secondly, from the experiential fact that no one is self-sufficient but is indebted to others for his very being and for his continued existence. In view of this social dependence it must follow *a fortiori* it [social dependence] extends also to the accomplishment of his final end, since his social obligations are also means to that end.

Secondly, the social group in its relation to God: just as the dependence of the individual to God is true, all the more true is the dependence of the social group to God. And, just as the spiritual being must elicit acts of religion [relationship] in accord with its nature, namely, spiritual acts, so must society by its very external [corporeal] nature elicit external [physical] acts of religion. Further, each man by the same law of his external nature, and by his participation in society, must share in society's external act of religion. Thus, man's need to express his religion externally is intrinsic to his nature. Over and above this, God has by positive decree determined the form this external manifestation should take on, namely, <u>sacrifice</u>, [the physical "pouring out" of self in the course of one's lifetime and the ultimate dissolution of the body in death. The necessity of sacrifice is encoded in Natural Law (genetic coding)]. (18).

Acts to be included in the supreme act of religion: men of all times have felt the universal conviction that they must make some offering to God. (19). They realize His supreme power in the contingencies that affect their daily lives; they acknowledge His supremacy by <u>adoring</u> Him. At the same time they are conscious of having offended Him and wish to <u>appease</u> Him. They are <u>thankful</u> to Him for having given to them life and other blessings; and they <u>petition</u> Him in the hope of obtaining future blessings. In these four acts man's needs in his relationship to God are realized, hence, they must be contained in the supreme act of religion. At the same time, this act being offered to the Infinite Deity must be proportionate to the Person to whom it is offered. Thus, the nature of sacrifice as fulfilling a reciprocal relationship between God and man will be considered.

Sacrifice. Sacrifice must be a legitimately instituted act. This means that the act has been fully approved both by God and society, and acknowledged as capable of satisfying their reciprocal relationship. On the part of man to God: it must express society's and each man's acknowledgement of the supreme dominion of God over them. Being an act of a visible body, it must be accomplished by means of a visible object, a victim, something the whole body can acknowledge as representing it, offered by [some] one[s] delegated. (This [these] duly delegated person[s] we will consider later.) Then, on the part of God to man: it must express God's satisfaction with *the object* offered as capable of establishing man in his relationship to God. And since the act is accomplished by means of an external object, the expression of God's acceptance of the object must also be made externally manifest in it. To preserve the reciprocal, social nature of sacrifice, this external expression of God's acceptance is considered an essential aspect of this supreme act of religion; otherwise, externally the act remains a one-sided affair, an offering on the part of the social body with no assurance of God's response to it. How then, is this external manifestation of God's acceptance to be accomplished? We answer: God Himself has determined by positive law the institution [natural] of sacrifice, both as to the objects offered and as to ones competent in offering it. Thus, if His directives are carried out in this act of religion, it is infallibly acceptable to Him. This acceptableness of the sacrifice to God, in order to preserve the social, reciprocal aspect of it, must take on a perceptible nature, which is accomplished by an apparent, external effect upon the object offered, i.e., destruction or the equivalent. Thus, the so-called "immutatio in peius" (change for the worse) would seem to be an essential part of sacrifice if its social, reciprocal nature is to be indicated in the act itself. (20). At the same time, since sacrifice by its very nature is the noblest act man can perform, (21) it also follows that the object which is offered is given a nobler status (22) by the function it performs, a status excelling that of all other material objects. Therefore, the "immutatio in melius" (change for the better) of the object is also, as a necessary consequence, realized, in virtue of the transcending purpose for which the object is destined. From all that has been thus far said about sacrifice, we can now derive a definition of it: *sacrifice is the offering, in a legitimate manner, of an external object acknowledging the supreme dominion of God.* The word "legitimate" indicates: the offering performed by one acceptable to God and man; that the object offered is representative of the social body and offered by them as such; that the same object is acceptable to God (23), which is manifested by an external effect upon the object, i.e., destruction or some other essentially equivalent change. The words "acknowledging the supreme dominion of God" determine it the supreme act of adoration, including within it acts of thanksgiving, propitiation and petition.

Fulfillment of Redemption. Thus far we have not spoken of the "One" competent to offer sacrifice. He must [be] represent[ative of] all, hence, he must be a man.[ahem, "human"]. At the same time, man has freely offended his Creator, an Infinite Being, hence, man is himself incapable of righting the wrong he has committed (24). Thus, it was God's plan that his Divine Son, true God in every respect, should take human flesh and redeem man from his fallen state. Man consequently, in virtue of this divine plan, takes on a whole new relationship with God. Man in his nature became sole mediator of spirit and matter. But in a more wonderful manner, Christ by taking human flesh became the sole Mediator between the Creator and His creation. (I Tim II, 5). Here in a far more excellent degree is the continuity of creation realized, where it has achieved [attained the status of] participation in the nature [divinity] of the Creator. (25).

Christ the Redeemer. All sacrifices of the Old Law were efficacious only through faith in Christ the promised Redeemer. (26). Thus, all priesthood of the old Law received competency [before] God through the priesthood of His Divine Son, since Christ alone as God and man, possessed both the authority and competency of offering to God man's supreme act of religion. Thus, in the foreseen merits of Christ's sacrifice, as priest and victim, did other sacrifices become efficacious for man[kind]. We find therefore, that the character of Christ *as sole priest among men* is intrinsic to the very nature of His being, and that [mark of character came about] from the very instant of His conception. (27).

Redemption Applied to Each. Just as the merits of Christ's sacrifice were already applied before it was accomplished, i.e., through the sacrifices of the Old Law, so there is need that in the New Law these merits be applied to each person. For this, Christ established His Church, His Mystical Body, perpetuated in visible form, in which "from the rising of the sun to its setting there is offered to my name a clean oblation, says the Lord of hosts." (Mal I: 11). Calvary is evidently Christ's consummated sacrifice, "sacrifice" in the proper sense as we have described it, by which man's status with God is completely righted. Still, each man [person] must draw for [her] himself from the abundant merits of Christ to effect his own personal justification. For this need, Christ established the Eucharistic Sacrifice, in which each can partake of the Sacrifice of Calvary. The Holy Eucharist too is sacrifice in the strict sense. (28).

Thus, through the sacrifice of Christ's Body and Blood is attained in a most perfect manner that divinely intended relationship between God and man. This then, establishes man, in the present economy of salvation, in his full relationship with God.

Our approach to the subject [of religion] was intended for the most part to remain on the rational plane. Thus, as to historical facts, e.g., whether through history the facts show that the destruction of the object of sacrifice is also present and thus establishes it as an essential part, has not been touched upon. Also, "revealed truths" have simply been presumed [taken on faith]. We have attempted to consider the *facts* as our faith informs us that they are, in the light of reason [faith derived].

But, the affect of our *rational consideration* of religion should [remain open to growth of consciousness], else it is really quite [ineffective]. [Knowledge changes, insights expand, and faith understandings change with informed vision.] If our intellects grasp this, our [changing understanding of] relationship to God, our wills should accordingly be ready to respond to the results of our intellect's searching. God grant [us openness, so] that our wills [are] found not wanting when [our intellects grow in understanding of rightly ordered and informed] relationship with the Supreme Good!

References:

1.) Gihr, <u>The Holy Sacrifice of the Mass</u>, Book I, Ch. I, footnote #2, pg. 17.

2.) Id. Book I, Ch. I No. 1, pg. 18

3.) Noldin, <u>De Principiis</u>, Rauch Innsbruck, 1953, Liber V, Quaestio Prima, No. 257; also footnote #2

4.) Gihr, op. cit. pg. 22

5.) Smith, <u>The Teaching of the Catholic Church</u>, New York, MacMillan, 1956, Vol. I, pp 265, 266

6.) Id. Pp 262, 273, 277

7.) Id. Pg 277

8.) Id. Pp 278, 279

9.) Id. Pg 296

10.) Gihr, op. cit., #5a, pg 24,

11.) Id. #5c, pg 25

12.) Smith, op. cit., Pp 42, 43

13.) LeFrois, "The Forbidden Fruit", <u>The American Ecclesiastical Review</u>, Vol. CXXXVI, No. 3, March 1957

14.) Smith, op. cit., pg 322 "Original State of the First Parents"

15.) Gihr, op. cit., #2, pg 60

16.) Id. Pg 22

17.) Id. #5d, pg 25

18.) Saint Thomas Aquinas: 2,2, q. 85, a 1 ad 1. "oblatio sacrificii in communi est de lege naturali; sed determinatio sacrificiorum est ex insitutione humana vel divina."

   Council of Trent: Session 22, Ch. 1. "Many theologians assert that sacrifice is strictly required by the very law of nature...Other [theologian]s do not grant this, but say that sacrifice is only in an eminent degree in accord with the law of nature." (Gihr, op. cit., pg 31)

19.) Gihr, op. cit., pg 29, also, Smith, op. cit., pg 88

20.) Generally, authors who maintain that the "immutatio in peius" is essential to sacrifice (presuming they base their reasons on historical fact, with the consequent explanation as to the reason "why" it is so), do so for the reason that by it man must acknowledge the supreme dominion of God over him, and shows God his willingness to offer Him his very life, which he does by the destruction of the victim offered, since apart from the social aspect we have considered, seems insufficient to prove it is an essential part of sacrifice.

21.) Gihr, op. cit., pg 31

22.) Id. Pg 25

23.) There are those who do not consider acceptance of the object as an essential aspect. Referring to Christ's sacrifice, Father D'Arcy, S.J., says, "The acceptance...is the complement of it and is not intrinsic to it." Cf: Smith, op. cit., pg. 511. Still, if sacrifice is not acceptable to God it is worse than useless, it becomes distasteful to God, as the prophet Malachy tells the priests of Israel. Mal: I

24.) Not that God could not have arranged other wise, the fact is that He willed condign satisfaction; not only were full amends made for man's sin, but man received a new and nobler relationship with God. (Smith, op. cit., pg. 509)

25.) In virtue of this continuity it would seem, as Scotus maintains, that even had man not sinned, still God would have arranged that human nature should be raised to partake in the divine; consummating in this act the noblest aspiration any creature could have. (Smith, op. cit., pp 492, 493)

26.) Gihr, op. cit., pg 35

27.) Suhard, E. Cardinal, <u>Priests Among Men</u>, Fides, Pub. Co., Chicago, pp 8, 9

"The Word, Who at once perfectly reflects the Father and is exemplar of creation, cannot, once He becomes incarnate, help but be the Mediator, the religious tie between God and man, and consequently, The Priest." (Salet, S.J., G., "Le Christ notre Vie", Casterman, 1937, pg 53)

"The priesthood, being a public function, legitimately belongs only to the one who is mandated by God, who receives official investiture from Him. Christ's priesthood is not an exception to that rule. The unique person of Christ ever hears the eternal words of the divine generation, 'Thou art my Son'. That divine decree simultaneously constitutes Him Mediator between God and man. Therein lies the metaphysical root of Christ's priesthood, its eternal foundation." (Bonsirven, <u>Epitre aux Hebreux</u>, pp 40, 41 and 267)

Thus, Cardinal Suhard, drawing from the teachings of the Fathers states: "Our Lord's Ordination was neither virtual nor inferential but proceeded from his double nature; 'The unction by which Jesus Christ was consecrated Sovereign Priest was the very divinity which filled and sanctified His sacred humanity at the very instant of the Incarnation'."

28.) It may be objected: if the Eucharistic sacrifice is constituted at the moment of consecration how is the visible nature of God's acceptance signified in the object offered, namely, in destruction? We answer: Man attains knowledge in a twofold manner, either through the evidence of the senses (experience) or through the evidence of others testimony (faith). As to the certitude of knowledge, that based on faith is the firmer because of the authority of the one testifying, God Himself. Men of the Old Law for the most part relied on he evidence of their senses, less perfect knowledge, but God was willing to accommodate them. In the order of the New Law: it is through faith in God that His redeemed people are to attain the supernatural life. Hence, to the Catholic believing in the Eucharist, no knowledge is more certain than the fact that bread is changed into Christ's Body, and wine into His Blood. The substance of bread and wine are no longer present, an essential change has occurred, no less real though imperceptible. Under the consecrated species is present the Victim of the Eucharistic Sacrifice. But, the external change must be found in the Victim of the sacrifice, not simply in the bread and wine or the species. This external change is signified (vi verborum) by the double consecration, by which the sacrifice of Christ's life is represented, i.e., His Body separate from His Blood. Thus, man's knowledge of the acceptableness of the sacrifice to God is fully satisfied and externally represented in the Victim; hence, the social, reciprocal aspect of the Eucharistic sacrifice is also realized. (cf: Smith, op. cit., pp 897-899)

Appendix B.

1978 Final Report. White House Conference.
Sylvester L. Steffen (HARVESTALL)
Proposals For Balanced National Economic Development.

Strengthening Local Economies

Local economies are best strengthened when local communities are self-sufficient. That is, when people discover that the solution to their needs are within their community, so that their dependency on the outside is reduced, be it for food or energy, be it for social needs.

Government can help by working with churches, schools, unions, business, etc., to educate to this mentality. Further, government can work to find ways, in which communities and individuals within them can become less dependent, viz., by teaching better land use, better basic nutrition, better energy utilization, energy conservation and energy collection, production and distribution. Government will do well to eliminate welfare programs, which promote a mentality of dependency, and will do well to show how greater independency can be achieved.

Since basic human needs, economic and social, are common to all people, effort in this direction will have the broadest possible application and serve the greatest common good, both for immediate and long-range economic solutions.

People and Jobs

Inflation, unemployment, growing tax burdens require root solutions, i.e., understanding on the part of all people that there are not outside solutions to local problems and needs, and that increased governmental involvement aggravates the problem. Structures, be they church, school, government, business or union, tend to become an end unto themselves and lose sight of their service responsibility, and in so doing create a dependence and subservience on the part of members that destroy personal pride, initiative and self-confidence to resolve needs. These institutions themselves become counter purposeful and insidiously subvert people so they become less imaginative, less productive and more dependent, not realizing that they must first contribute to the structure before it can contribute to them, and that in the exchange more can be lost than gained. Bureaucratic proliferation is a major factor in uncontrolled inflation and will lead to inevitable bankruptcy. Government can specifically work with churches, schools, businesses and unions to assist them in their common service effort to help people discover solutions to community needs, including health and welfare, and best utilization of resources they have at hand. Such a nationwide effort might be identified as COMMUNITY SUFFICIENCY or PROJECT UNWIND, i.e., freeing communities from wasteful bureaucratic red tape. Such an approach gets to root solutions in local communities and does not perpetuate arbitrary and wasteful creation of jobs by government, which, short-range may seem plausible in some communities, but not in others or over the long pull.

Government Budgets

The closer resources can be kept to where they are needed, the greater economies result. Government, both State and Federal, should seek to reduce their involvement in people's lives and help people to discover how to find answers to their needs in their own community so that community resources are not drained by growing tax burdens, which come back as a mere fraction of their original amount if at all.

Geography of Growth

Agriculture, i.e., tendering food from earth, is the umbilical that supports all human life. All people should be encouraged to have some involvement with earth so as to optimize its immediate resources and minimize people dependency. Ways of recovering energy from the immediate environment should be shown. Again, government influence should be through existing local structures, church, school, business and unions, helping the people discover and employ their community resources to solve community needs.

Government and the Management of Growth.

The common needs of all men emphasize the commonness of interest and the need for individual interest (group) not to function at the expense of the common interest. Again, this social mentality has to exist amongst people as part of their value system before they can become selfless enough not to make personal demands at the expense of others. Business cannot be allowed to strangle communities because of peoples' dependence on them, e.g., energy marketers. Government should not come in with its own structure to "impose" on a community, but should help people discover a sense of community, which is sensitive to all people, even beyond their immediate community so as to support communities, which may be lacking in essential resources. Any additional structure should grow out of existing community structures of church, school, business and union.

Streamlining Government

Government should take the lead and realize its sense of service to the people, and that when it becomes a burden to people it has lost its sense of service. The present direction of government is to inevitable greater public catastrophe because of its economic strangling of the public through waste and growing tax demands. Its damaging impact is twofold, i.e., draining the economic resources of people in a pyramiding spiral, and encouraging a public mentality, which makes people more and more dependent on bureaucracies. Inevitably this leads to the destruction of democracy because of bankruptcy resulting from people demanding more from government than government's resources allow, and government will be forced to use force to maintain control over the populace and prevent anarchy. Efficiency must first be found within government before government can help the community to become more efficient.

All facets in the lives of human beings must integrate the whole person, the whole community. To perpetuate a notion that religion, politics and business have their own ethic, which cannot be mixed, is incorrect, because they are all aspects of people living together. Religion conveys a value system, which inevitably influences individuals in their private and public lives. If the dogma of religion is strictly other world oriented it can lose credibility in this world, for the here and now is the only world where it exerts obvious influence on peoples' lives.

To put government, religion, school, business and unions in a competitive attitude toward each other, for whatever reason, is divisive and destructive of community and drains resources. Survival of human society on earth demands an end to this gaming mentality and requires greater common effort on the part of all human structures. Government can be a powerful influence to this end and must be, for <u>government is people functioning in rational ways for the common well-being.</u>

July 1978 Final Report
White House Conference
On Balanced National Growth & Economic Development
Appendix Volume 4
Pp. 100-102

Superintendent of Documents
U.S. Government Printing Office
Washington, D.C. 20402

Stock No. 052-003-00544-1

Appendix C.

1984 Cereal Chemistry Seminar:
"Commercial Value of Live Grain"
[The Water Factor]

October 18, 1984
Sylvester L. Steffen

North Dakota State University
Department of Cereal Chemistry
Dr. Orville Banasik, PhD.
Dr. Joel Dick, PhD.
Harris Hall, Room 12
Fargo, ND. 58102

The Biological Utilization of Physical Energy

1. Connections
2. Atom to Adam
3. Seeds and Insights

Commercial Value: The Water Factor
1. Dead Grain vs. Live
2. Photoelectric After-Ripening
3. The Case for Interdisciplinary Science

Sylvester L. Steffen

The Biological Utilization of Physical Energy

1. Connections.

Rodney Dangerfield has asked perhaps the most critical question of today, "Why ain't I got no respect?" The white-hot political issue is respect for life—the moral dilemma of individual right to life, of the common right of all life and of unborn future life. The moral dimensions of human overcrowding, of ecological poisoning, of depletion of resources, waste of interdependent life, of war, of pestilence and hunger, are not seen as right-to-life issues. The even more basic question, "What social bias has caused the cultural ingraining of a public attitude that causes these horrible abuses?" has hardly been raised.

Science and religion are in quest of consciousness of relationships that reveal and compel to right order, to the common good. Universal religion is consciousness of relationships in the universe, and the discipline of this awareness on personal conduct. Science and religion are the same in their common objective, truth. Disrespect for truth, whether by science or religion causes disrespect for life and for person.

The path from chaos to consciousness in the universe is the history of relationships, of the interaction and unfolding of energy in matter, physically and biologically. Consideration of this is of practical and academic purpose. The path of conscious chaos is a human defect.

A workable definition of intelligence is "consciousness of purpose". A workable definition of morality is "conscious conformity to universal purpose". There is no morality in ignorance (!), which states the moral imperative compelling every individual to inform personal intelligence. The imperative of personal morality is an imperative of science. Science (knowledge) is the conscience of religion. When speculative knowledge takes liberties with truth, science proves or disproves with evidence. Religious speculation in incompletely understood relationships is at risk of erring and of being exposed. The existing body of collective science is so vast that laws compelling to universal religion should be obvious.

God's law of universal relationship of life is indelibly written in water. Divine sensitivity is in the transparency of water. The providence of life (food-energy) is in the working of water, which rearranges matter in the ascent of life on its spiral staircase, DNA. This incredible structure is the operative mechanism of life, and is its library. Significant adjustments in the history of life's unfolding are recorded there, and memory machinery imprint experiences for future use. DNA is the computer chip by which Worthy Purpose (sacra mens) functions "by nature" and "by nurture" to adapt life to physical and cultural circumstances along its way. Poisoning ecology, wasting life and depleting resources are violations of the Worthy Purpose of nature and are unethical acts. If religion is authentic it must speak out against these sins as strongly as it does against abortion.

In the debate of vital issues, too much energy is expended in producing too much heat and too little light. Pointless expenditure of energy causes entropy. Violation of sacrament produces entropy. De-molecularizing and de-atomizing matter return it to uselessness for life. The Worthy Purpose of life is accomplished in molecularizing, in the controlled entrapment and utilization of energy in matter. Sacrament is both the motive [objective] and motivation of life. Sacrament of food (eucharist) comes from sacrament of water (baptism). In the cycles of life energy is purposely engaged in the furtherance of sacrament, that is, in the advancing of life in matter. Life is itself the expression of energy in highly organized matter, whether, the human body molecule, or whether the less grand grass seed.

When least life is wasted, all life is diminished. Disrespect for life is so prevalent because a consciousness of connections is missing. Humans fail to see the connectedness of matter and energy in their common affairs of daily life, of agriculture, of science, of religion, of education, of politics, of business, of economics. All of these contribute to the seamless mantle of life enveloping earth, and because of human insensitivity, all life is threatened with catastrophe.

Defining terms. [The words] Heat and energy are commonly interchanged. But, they are used incorrectly. The energy-state of seeds is decreased by exposure to heat. The application and expenditure of energy are evidenced by heat. Heat is the "feel" of an object, a measurable condition of molecular friction, that is, [of] the speeding up or the slowing down of electron rotation. It is a form of communication. Temperature is the language communicating the message of radiating heat. Energy is potential for work. It is built up in matter by converting heat radiation to chemical energy. The common source of radiant heat is the sun which floods earth with biologically useful waves in the infrared and visible light spectra.

2. Atom to Adam.

A systematized understanding of physical order in the universe surely is in understanding operative laws in the organization of subatomic particles into inductive/repulsive machines, atoms and molecules, electric motors.

It is presumed that the formation of all matter operative in the universe is from the accommodation of forces of attraction/repulsion associated with subatomic particles and their bonding under energy equilibrium of centripetal and centrifugal forces. Protons constitute the nucleus of atoms and have a positive, centripetal potential, while electrons, under the influence of, but outside the nucleus, offset the energy of the nucleus with an essentially equal but opposite (negative) centrifugal potential. Simplistically, this characterizes energized matter accessible for structuring in life.

Physical interactions of atoms occur by which electrons of one atom come to be shared with another atom, and, by which one atom will give up an electron and another will accept the electron. Presumably, pathways of this sharing and surrendering of electrons may involve even subatomic components of electrons. Subatomic particles, electrons and atoms involved in this giving and taking, and having positive and negative charges are "ions". An atom that wants to accept an electron is positively charged (cat-ion), and one that wants to give up an electron is negatively charged (an-ion). Light itself is a stream of particles called "photons" whose individual quantum of energy varies inversely with the wavelength. Photons are energy agents intercepted in chlorophyll and used in assembly of glucose (6)C-HOH, and are trapped by photosynthesis in the molecularizing of life. The action of photons in plastids (chlorophyll) is at the electron and/or subatomic level, and may be operative in visible and non-visible light spectra. Its contribution is presumably quantitative.

In view of the inductive/repulsive potential of centripetal and centrifugal energy in every atom and molecule, each is an engine in its own right with a potential for work. The assembly of these machines has been in the foundry of the cosmos over billions of years, and their unique construction is the handiwork of operative forces in space, gravitation, radiation, etc. They possess the ability of responding to solar radiation, for example, by having the dynamic equilibrium of subatomic constituents raised to more increased motion, and coming to a higher energy-state. This responsiveness is selective, that is, molecules (atom) of a particular character will respond to wavelengths in specific bands. For example, water and carbon dioxide are particularly responsive to specific wavelengths in the infrared band. The increased energy potential is from attenuation of harmonic (sympathetic) wave energy.

Laws governing energy and matter are highly predictable and reliable. It is this redundancy that puts matter and energy in relationship and which puts organized structures in cyclical relationship. This reliable redundancy of nature is what secures faith in its providence. The determination of this redundancy is [natural] "purpose". By virtue of its beneficial consequences, the containment of energy in atoms and molecules for purposeful work is "holy". The word for "holy purpose" in religion's language is sacrament. Sacrament is centripetal energy [syntropy]. The opposite of sacrament is entropy; by it, energy is let free to escape from molecularizing. Entropy is centrifugal energy. Wasting life is entropy. Destroying seed life is entropy. Denuding rainforests is entropy. Chemical agriculture is entropy. Farming marginal land is entropy. Spoiling water is entropy. Draining wetlands is entropy. Paving topsoil is entropy. Removing nuclear energy from atoms is entropy. All of these violate [natural] sacrament. Entropy destabilizes energy equilibrium accomplished by molecular arrangements of life, and hastens chilling of earth by disintegration of its [her] seamless mantle. Use of energy must be with conscience, as must care to replenish it for future generations.

The continuum of life, characterized by its obvious purpose in cyclical connectedness in water, confirms unity of life, the more so that DNA is substantively identical in all living cells. There is no biochemical activity in cells except with relation to water, which is both its medium and principle contributor. This awareness is deep-rooted in human consciousness and intimately associated with religion. It speaks not of human separation from the order of life, but to dependency within it. It speaks not of kingly privilege for humans, but of responsible stewardship and kinship. It speaks not of disparate creation of various life forms (creationism), but of evolution by symbiosis from common origins.

The bankruptcy of literal, creationist theology is evident in its legacy of ecological disaster. The postponement of entering into a [n enlightened] New Testament consciousness of living in symbiotic relationship, and of conscious respect for all life, can no longer be allowed. Religion must update its scientific consciousness to contemporary knowledge of relationship, and science must preserve its ethical responsibility of being faithful to truth.

3. Seeds and Insights.

The destiny of a live plant is wrapped up in the production of seeds, and the destiny of the seed is to produce a live plant. In this cyclical order, energy and matter come to be used over and over again. But more than that, subtle additions that escape perception happen. Increase [by the agency of water] is given to living matter.

The seamless mantle of life on earth, that is, ocean-life, the soil, all flora and fauna and the atmosphere, is that significant addition, and the components of it, have their origin in the successful entrapment of photons coming from the sun. Photon energy is particular so that entrapment of it makes a substantive contribution to living earth.

Biochemical utilization of solar energy is at the electron and sub-electron level. Coming to knowledge of the mechanisms and processes involved is to come to greater understanding of the origin of life itself. The gaseous mixture of the atmosphere, minerals of earth and oceans of HOH are the organized substances used by life to steal sunshine and to raise its energy to consciousness in the human body molecule. Of this order all life is beneficiary, and the reciprocally responsible benefactor to it.

With humility and respect humans should seek greater insight into the order of nature, and use their insight to symbiotically facilitate its Worthy Purpose. Here is the cathedral for authentic worship. Sacrament is inescapably the moral obligation of science and religion because of human consciousness. From communication comes consciousness, and from consciousness, conscience.

Our purpose here is to increase our understanding of the commerce of life in seeds, and from this consciousness make symbiotic connections to real [everyday] life, also for greater personal respect for life. Einstein's [energy/matter equation] deals directly with the economy of energy on earth, substantive effects on matter, and specifically, with the physics of photo-electricity by which sunlight energy restores what is lost to entropy. The critical link is the seed. Without the seed there is no plant, no photo-electric reserve of food, and only limited life on earth.

Commercial Value: [The Water Factor].

1. Dead Grain vs. Live.

Life is a continuum of successive processes, which provide for growth and maintenance of organisms, and is a value worth preserving also in commercial grain. This scientific observation seems innocuous enough, but it has instigated much controversy and adversary response in the grain drying industry for HARVESTALL's commercial commitment to it.

Dr. Edward Deckard, Agronomist at North Dakota State University is quoted in a 1977 trade publication: "I know of no published data showing germination percentage to reflect nutritional quality of grain." (3). Relating to this view is that of H. S. McKee who believed an observation of Vasiliev in 1908 was still valid in 1949: "It is remarkable that up to the present time seed ripening has been so little studied, though the process is of the highest interest as we are here dealing essentially with the synthesis of organic substances". (1).

I believe the observations [of Vasiliev] are still true except for my work and the commercial commitment of HARVESTALL since 1960. If there are more recent findings, they have not influenced industry. The unpublished Master of Science Thesis of this writer, Effects of Drying Method on the Germination of Corn (4) addresses the relationship of germination to commercial value of grain. Steffen Patent 3,408,747 (November 5, 1968) specifies maintaining the environment of stored seeds so as to accomplish temperature/moisture equilibrium with atmospheric air, with low inputs of heat at most, and the use of levels of ventilation as determined by volume and wetness of grain so as to preserve intrinsic biological values, including seed-life. This pioneer process patent and others that have followed it are controlling in natural-air/low-temperature drying and atmospheric ventilation of [life-secured] grain.

Aggressive marketing under these low-temperature-drying patents since the mid-sixties has had widespread market impact, which continues to the present time. The presumption of HARVESTALL Marketing was that the authenticity of HARVETALL science would be reinforced by user experience [and it has been], and that continuing research at land grant universities would document devaluation of grain from exposure to heat, and [that universities] would eventually verify and endorse the seed-science of HARVESTALL. Unexpectedly, agricultural engineers at land grant universities responded adversarially to HARVESTALL and engaged the Extension Service network in their public aggression. And with catastrophic damage to HARVESTALL, I might add.

HARVESTALL believes that its market effort is sounded in objective science [and] with consideration of the multiple sciences involved; that its teachings and their application are in the public interest and particularly, in farmers' interest. HARVESTALL welcomes good faith, critical scrutiny of its science and conduct.

This backgrounds the reason for my presence here today. I am personally indebted to Dr. Orville Banasik and Dr. Joel Dick, of the Cereal Chemistry Department, for the privilege of presenting my thesis.

Simply stated, the thesis of HARVESTALL is:

THE ELECTROLYTIC POTENTIAL OF LIVING SEEDS DIRECTLY DERIVES FROM ACTIVE EFFECTS OF SOLAR RADIATION IN THE VISIBLE LIGHT AND INFRARED SPECTRA, IN THE MEDIUM OF CELL WATER. THE INCREASED ELECTROLYTIC POTENTIAL EXPERIENCED IN CHILLCURED CORN IS FROM GREATER ACTIVE SOLAR ENERGY INPUT DURING THE RIPENING PROCESS. THE BENEFICIAL EFFECT OF ACTIVE COLLECTION, USE AND STORAGE OF SOLAR ENERGY IS IN MAKING THE SEED A MORE USEABLE FOOD RESOURCE, BOTH QUANTITATIVELY AND QUALITATIVELY, WHICH IS ANALOGOUS TO CHARGING A BATTERY TO ITS FULL POTENTIAL. FULL COMMERCIAL VALUE OF THIS FOOD POTENTIAL IS ACHIEVABLE ONLY IN LIVING GRAIN, NOT IN DEAD AND DAMAGED GRAIN, AND ONLY IN THE SUCCESSFUL COMPLETION OF RIPENING PROCESSES.

2. Photo-Electric After-Ripening.

A commonly experienced frustration in cold weather is to step into one's car, turn on the ignition, get several painful groans, and then nothing. Dead battery. Unless seeds are allowed to charge their batteries, the embryo may experience something of the same frustration when it wants to grow. The best analogy I can think of to illustrate "after-ripening" in seeds is that of "charging the battery". Alternately, this ripening process is properly termed "natural CURING". A battery is charged when a maximum reserve of electron energy is stored in chemical form. The charging of the battery is accomplished by a specific mechanism and process that brings its [battery's] chemistry to an electron-rich state. In the automobile it is the alternator; in the seed (cells) it is the mitochondria and plastids.

All physiologically active cells typically have both mitochondria and plastids. The common fuel produced by them, and used to power all living processes of cells and living systems is ATP (adenosine tri-phosphate), as phosphate sugar derived from glucose, C-HOH. In the case of mitochondria, the production of ATP is by oxidative phosphorylation (as in the Krebs cycle), whereas, in the case of plastids it is anaerobic and by photo-phosphorylation, and powered by solar radiation whose energy is molecularly accessible by way of water and carbon dioxide. Photon energy is the operative energy.

To understand where and how biochemical processes work in seeds, a review of seed morphology is in order. The seed of *zea maize* is representative of grass seeds and convenient to our discussion.

The most obvious structures of the corn seed are the embryo (germ) and the endosperm. The embryo is a shield-shaped structure that is called by the Latin name scutellum. <u>Mitochondria are present in the cells of the scutellum but not in endosperm</u>. Plastids are more predominantly present in the cells of the endosperm. Germination and growth of the seed are powered by oxidative phosphorylation in germ cells. Upon hydration [seed uptake of water], gibberellins are produced by germ cells and cause aleurone-protein to be hydrolyzed. The aleurone is a structure of protein-packed cells wrapped like a skin around the endosperm, just under the seed coat. With germination, enzymes act on the endosperm starch to make it accessible food for the incipient plant.

Photolysis and phosphorolysis are presumably activated in water. Wetting of grain by soaking (hydration) and wetting by hygroscopic absorption of moisture from the air do not necessarily have the same effects. Absorption from atmospheric air temporarily causes weight reduction, presumably from increased rate of respiration, yet, the long-term effect may be of weight gain. In flour made from seeds, the nutritional utilization of food substances, minerals, etc., is presumably enhanced by photolysis, phosphorolysis and hydrolysis. The addition of yeast to dough produces carbon dioxide by fermentation and increases acidity, presumably with photophosphoro-hydrolytic responses, for example, making phosphorus, magnesium and calcium accessible from [seed] phytin.

In the formation of most seeds the endosperm is short-lived. Cereal grains and several other species differ characteristically in the transformation of the endosperm into a major organ of stored food.

Typically, crystalline starch and protein are accumulated in plastids. Endosperm cells abound in plastids laced with crystals of modified starch. The characteristic yellow of the endosperm is of the plastids. Plastids are the material of chlorophyll and are made up of striated layers of grana, battery-plates, if you will, holding charges of electron energy.

Since seeds carry self-contained food supplies they must contain ATP and/or materials for its production. Original derivation of newly synthesized ATP is with materials of photosynthesis and occurs in chloroplast. Nearly one-third the structure of chlorophyll is composed of phytol, an alcohol that is highly reactant with oxygen. Phosphorus is an essential mineral intimately associated with plastids, but is typically present in seeds in a poorly accessible form, phytin. Phytin is a calcium magnesium salt of inositol phosphoric acid. It is known that availability of phosphorus improves in grain with storage.

With the onset of senescence, leaf and stem cells deplete themselves of food materials, even proteins, which must be hydrolyzed to amino acids before they can be translocated [to the seeds]. As this is taking place, waving fields of grain change from green to gold. A large proportion of phosphorus of mature plants ends up in seeds and is associated with plastids of the endosperm cells. Yellow pigmentation reveals carotenoid structures typically associated with chlorophyll in grana. In view of absence of mitochondria in the endosperm, the transformation of materials in the endosperm is presumably by the photo-responsive action associated with grana. Cell walls of the endosperm may degrade to facilitate biochemical response.

The production and accumulation of starch and protein in plastids are interactive, photosynthetic processes. Protein is simply modified starch, 85% C-HOH. The accumulation of starch and protein in cells is by way of phosphorylation, the Krebs cycle. Presumably, plastids continue to play a key role both in fueling enzyme-controlled processes and in storing accessible electron energy, ATP.

Protein of the aleurone [consists of] large molecules [made up] of many glycosidic linkages. Obviously, presence of protein in the aleurone is by way of being constructed there. It is likely that in the transformation of starch and protein in plastids of endosperm cells, transfer of amino acids from inner cells to outer cells provides material to structure the protein of the aleurone. This is presumably a significant process of after-ripening and one of commercial importance. Carbon dioxide and water are critical contributors. Removing water by artificial drying and denaturing protein with heat are gross catastrophes to the seed for [the] conclusive inhibition of after-ripening processes.

Maintaining an optimum condition for cells of the endosperm to promote photophosphorylation is by way of hygroscopic interaction with the atmosphere. In that protein formation is slow, particularly under cool temperatures of late season, it is important that a continually favorable environment be preserved throughout storage. This means holding seeds to optimum moisture and temperature. In practice this may best be accomplished with seasonal wet-bulb temperatures, which regulate both seed-temperature and seed-moisture. The interaction of temperature/moisture aptly defines both a "dormancy index" for grain, and conditions leading to chemical stability for seed food...

The particular responsiveness of water and carbon dioxide to infrared [radiation] by [the] attenuation of energy in specific wavelengths is believed operative to photophosphorolysis here. Ventilation of grain with atmospheric air, and [the] use of infrared radiation [electrical heat lamps] to condition air-humidity [energize it] may contribute significantly to [the] photophosphorylation of ripening seeds in storage, to aleurone [protein] synthesis, to ATP accumulation, to greater electron potential, to [the] fixing of hydrogen and oxygen in [seed] carbohydrates, etc.

What factual evidence exists to support the speculative mechanisms and processes suggested? There are data showing significantly increased rates of emergence in corn [seeds] after 18 months in storage as compared with rates of emergence after 6 months storage, 83% and 56%, respectively. (5). The modification of food reserves [and the] increased electrolytic potential over storage are postulated causes, namely, more abundant reserves of electron-energy, more accessible ATP, for example. There was no significant difference in germination percentages of corn stored below 18% moisture. Improved emergence was greater at higher moistures.

Data from blending dry corn (9.9% moisture) with wet corn (18.8% moisture) showed an inverse linear relationship with the gradual reduction of seed moisture to increased dry weight, both in originally dry grain and in originally wet grain. After 240 days a 3% increase of dry weight over original dry weight was obtained in the wet grain. (Published in Grain Ecology.) (5).

But by far the most convincing data are those of thousands of on-farm practitioners of the CHILLCURING science of HARVESTALL since 1970. These have been reported in investigative articles of trade publications, particularly of the electrical power industry…[The] testimonial ad (Back to Good Grain) was the [adversarial] focus of the Minnesota Extension Service, Agronomy and Plant Genetics, "Crop News, No. 40", May 1978, *Weight Shrink and Dry Matter Change During Drying and Storing Corn Grain*, D.R. Hicks, H.A. Cloud, and L.L. Hardman:

"Summary…we conclude that a dry weight increase in the amount claimed by Harvestall in their "Chillcuring" process is not possible. In fact, the research evidence indicates that when shelled corn is subjected to the environmental conditions that exist in a Harvestall "Chillcuring" system there will be a dry weight decrease…"

Notwithstanding this adversary conclusion HARVESTALL stands by the authenticity of its science as documented in this testimonial ad. HARVESTALL [believes] Crop News 40 [is mistaken in its science and that its] intended purpose was to publicly discredit HARVESTALL's science and representations, and it has done so by asserting false presumptions…

[The Minnesota Extension Service turned HARVESTALL claims over to the Minnesota Attorney General for prosecution. I personally responded to the Minnesota AG's inquiry. The conclusion of the AG was that the issue was a dispute of science, which was not yet resolved. No action was taken against HARVESTALL, which continued its market claims and enjoyed phenomenal market acceptance.]

3. The Case for Interdisciplinary Science.

Given the now known degrading effects of heat on grain and high costs of energy, hot air drying of grain cannot be justified either scientifically or economically. That the land grant universities are largely responsible for causing widespread acceptance of kiln-drying [of corn grain] imposes on them the obligation of informing the public of the negative consequences of it. Neglect to do this is breach of science and breach of public trust.

How did universities ever get into this predicament? Surely it is consequential from failure to anticipate the sensitivities of seeds as biochemical organisms and criticalness of environment as it effects biochemical values. Chemistry, physiology, physics, atmospheric science and other scientific disciplines converge in the art of caring for grain [seeds]. The interaction of these disciplines has been overlooked, probably in part because of politics of governmental distribution of money and governmental assignment of public responsibility through the land grant system.

America's granary has been entrusted to unqualified keepers. Either the keeper must be able to become qualified, or else government should reassign this public trust to another keeper, perhaps to cereal chemists, to plant physiologists. With all due respect, agricultural engineers breach science and the public-trust in the pseudo-science they promote. Their late endorsement of low-temperature/natural-air drying has been by the force of market pressures. Neither the government not the public can condone their breach.

Erroneous presumptions inevitably give rise to pseudo-science. The presumption that seed-life is inconsequential to commercial value of grain is offensive even to a novice scientist, and is justifiable by no known standard of science. Much less is persistence in error excusable in light of decades of research and farmer experience, which documents devastation of grain by heat, inflationary waste of agricultural resources and loss of foreign markets because of degraded [grain] quality. Country elevators have become time bombs from dust of degraded corn. [Grain dust explosions and fires at country elevators are not uncommon.]

The facts of record suggest that more than oversight has been at work in view of unembarrassed public aggressions by agricultural engineering and extension service personnel against the teachings and marketing efforts of HARVESTALL. An obvious way is for land grant universities to restore public trust by being honest with science, even if it means endorsing [another's] science…

———————————————

These are still the "ChillCuring" benefits to grain and to farmers: "ChillCuring" allows farmers to field-shell corn while it is yet wet and uncured; "ChillCuring" ventilates the stored grain with controlled volumes of atmospheric air and optimizes the after-ripening of the seeds (by heat removal from the grain) even as free grain moisture is removed from the storage bin. Grain ripens to full food and weight values as it is stored, and the farmer uses free atmospheric air to dry it with least energy consumption.

The validation of Harvestall's seed science of "grain ChillCuring" is now a matter of public record. The pioneer seed science of Sylvester Steffen teaches "ever-normal" corn keeping; it is disclosed in product and process patents, U.S. Patents: 3,408,747; 4,045,878; 4,045,880; 4,053,991; 4,077,134; 4,148,147; 4,175,418; 4,247,989; 4,256,029; and 4,800,653; Canadian Patents: 1,086,052 and 1,090,562; and it has been validated by on-farm usage and by legal challenges in U.S. Federal Courts of Minnesota and Indiana, and in Iowa District Courts of Chickasaw, Jones and Linn Counties.

Historical note: Harvestall's products and technology were marketed from 1977 to 1982 by Harvestall Marketing, Inc., Trimont, MN, formerly, Persson Grain Company. Governmental interference, including, the disallowance of previously allowed energy tax credits for the qualifying "active solar" system, the discontinuation of the storage facility loan program to farmers, and the [USDA] University Extension Service's active endorsement of kiln drying and published discrediting of Harvestall "ChillCuring", all conspired to precipitate the business collapse of Harvestall Marketing, Inc, in the 1980s. The continuing depression of the agricultural economy thereafter has stymied the recovery of the grain storage business. The Harvestall ChillCuring System is not now being marketed.

Afterword

Ratio Mutans Fidem Mutandam
(*Faith must change with changing knowledge*)

THE SPIRITUAL QUESTING OF FAITH AND REASON is the prolix fertility of consciousness. *Trimorphic resonance*, the harmony of Reason, Faith and Purpose, is a resurrection process; it is the soul harmony of deep consciousness. New consciousness sets in motion new transformation; transformation is the deep purpose of quantum sexuality—love's purpose and elemental compulsion. To know is to love; to love is to live.

THE DEEP BONDING OF LIGHT is the beginning nuptial of knowledge. Life's first stirring is the deep sexual essence of vitality, the incipient necessity of perpetual resurrection. The photons, embedded in the electrons, weave the seamless and brilliant patterns of life, illumine every open intuition, and serve the purposes of stitching together the rational fabric of the cosmic continuum.

LIGHT INSISTS ON NEWNESS. When old fabric frays, it releases its light to the rainbow and returns its dust to Earth. The liberated photons and motes are given back for new assembly and new consciousness—the eternal dance of new becoming.

THE PRIVILEGED ASSEMBLIES OF VITALITY are called to celebrate their joined coming to light—the Way, the Truth and the Life. New consciousness is called to celebrate the new resurrection of knowledge and its amplification in Faith—a divine, eternal mission—divinely and humanly embedded in and amplified throughout the deep "Sea of Infinite Substance".

Consciousness is A Rising Tide

HARMONY MITIGATES VIOLENT POLARITY. In nature, polarities mitigate their difference and channel potential toward instructive purpose. The willful mutuality of Faith and Reason channels opposing potentials. Quantum Religion is Faith-informed Reason, the constructive agency of purposeful Word and Work.

OUR FATHER'S FAITH is a family inheritance of cosmic wisdom whose reasoned purpose sustains common vitality and community.

TIME REQUIRES EACH OF US to give back our light and our dust for the new generation. We are growing, maturing, and seed-producing plants, destined to rise and decline, and to arise again in new assemblies of light.

THE PERSONAL DISSOLUTION OF SELF in service to others can be a willing giving or a self-defeating struggle obsessing in self and things, and resisting the inevitable diminishment of the self.

PERSONAL SELF-FULFILLMENT is better served by non-violent good will, which also better serves global harmony and the uplift of communal consciousness.

THE "GOOD NEWS" IS PURPOSE the "rising consciousness" of communal intention that is born of Reason and Faith. Conscious purpose —conscience— intends the symbiotic accommodation of differences not rigid intolerance of them and social undoing. Reason/Faith/Purpose are Trinity Grace—the cohesive resonance of communal civility.

www.secondenlightenment.org

Selected References

Joseph Campbell, "Thou That Art, Transforming Religious Metaphor", edited by Eugene Kennedy, 2001, The Joseph Campbell Foundation, New Way Library, 14 Pamaron Way, Novato, CA 94949

Michael Crosby, "Do You Love Me: Jesus Questions the Church", 2000, Orbis Books, Maryknoll, NY 10545-0308

Charles Dickinson, "The Dialectical Development of Doctrine, A Methodological Proposal", 1999, printed by Pryor Pettingill, Inc., Ann Arbor, MI 48107

Bernard Evans & Gregg Cusack, Editors, "The Theology of Land", 1987, The Liturgical Press, Collegeville, Minnesota

Roger Haight, "JESUS Symbol of God", 2000, Orbis Books, Maryknoll, NY 10545-0308

John Haught, Editor, "Science and Religion, in Search of Cosmic Purpose", 2000, Georgetown University Press, Washington, DC

C. Hibbert, "The House of Medici: Its Rise & Fall", 1975, William Morrow & Company, Inc., New York

Kenneth Howell, "God's Two Books, Copernican Cosmology and Biblical Interpretation", 2002, University of Notre Dame Press, Notre Dame, IN 46556

J. Wentzel van Huyssteen, "The Shaping of Rationality", 1999, Wm B. Eerdmans Pub. Co., 255 Jefferson Ave. S. E., Grand Rapids, MI 49503

Ramon G. Mendoza, "The Acentric Labyrinth, Giordano Bruno's Prelude to Contemporary Cosmology", 1995, Element Books, Inc., Shaftesbury, Dorset SP7 8BP

Mary Settegast, "Mona Lisa's Moustache, Making Sense of a Dissolving World", 2001, An Alexandria Book, Phanes Press, P.O. Box 6114, Grand Rapids, MI 49516

Laura Sewall, "Sight and Sensibility", 1999, Jeremy Tarcher/Putnam, NY

Edward Shils, "The Virtue of Civility", edited by Steven Grosby, 1997, Liberty Fund, Inc., 8335 Allison Pointe Trail, Suite 300, Indianapolis, IN 46250-1684

C. Taylor, "Hegel", 1975, Cambridge Univ. Press, 40 W 20th St., New York, NY10011

David S. Toolan, "At Home in the Cosmos", 2001, Orbis Books, Maryknoll, NY 10545

Hugh Trevor-Roper, "Crisis of the Seventeenth Century – Religion, the Reformation and Social Change", 1967, The Liberty Fund, 8335 Allison Pointe Trail, Suite 300, Indianapolis, IN 46250-1648

Cletus Wessels, "The Holy Web, Church and the Universe Story", 2000, Orbis Books, Maryknoll, NY 10545-0308

BOOKS by Sylvester L. Steffen

PRIMARY SCRIPTURE: Cosmic Religion's First Lessons.
First Lessons premise the unity and identity of cosmic energy/matter as the spiritual/material origin of the essential substantiation and vitality of soul/body sacrament, the natural basis of the Christian Commandment "to love God and one's neighbor as oneself".
www.1stbooks.com, 2001

2 0 0 0: A Summary Prevision toward Global Revitalization.
The Prevision indicts the acculturated breach of trust as the root cause of frustrated consciousness, societal disharmony, ecological desecration, and the defeat of altruistic ascendancy (caritas).
Self-published, 1999

EDEN'S LIFEWORK POETREE: A Reconciliation of Science and Religion. (9 Chapbooks of poetry)
The Reconciliation identifies faith/hope/love as the rational outcomes of the resonant processes of communication, consciousness and conscience, and "sacrament" as the soul/substance edification (intensional/intentional) of cosmic energy/matter. The cumulative effects of failed reconciliation are subjects of essays and poems.
Self-published, 1998

About the Author

Author Sylvester Steffen – a lifetime student of theology and biology, business entrepeneur, and practitioner of family responsibilities – embodies a rich and differentiated mind of cross-disciplinary insight. His wide personal experience, his interests and study make him a communicator rich in urgent ideas. Steffen is steadfast in confronting institutional self-obsession, whether religious, political, scientific or other. His eclectic mix of worldly wisdom and moral discipline confronts cultural myopia, the self-interest focus of professional disciplines. Author Steffen claims no quick answers to complex social problems but he does advance a creative process of societal dialog by which cross-purposes might be reconciled with a common purpose. His interdisciplinary approach to rationality is enlightening for old theologies in need of updating, as well as for politics and economics fixated in feudalistic philosophies that exploit the marginalized and advance globally the politics of ecological disregard.

www.ingramcontent.com/pod-product-compliance
Lightning Source LLC
Chambersburg PA
CBHW031822170526
45157CB00001B/155